D1739253

The Call Goes Out

Messages from Earth's Cetaceans

Interspecies Communication

by

Dianne Robbins

Hidden Mysteries
TGS Publishers

The Call Goes Out: Messages from Earth's Cetaceans
by
Dianne Robbins

Published in 1997
Copyright © 1997 Dianne Robbins
All rights reserved. This book may not be reproduced, stored in a retrieval system, or transmitted in any form by electronic, video, laser, mechanical, photocopying, recording means or otherwise, in part or in whole, without written permission from the publisher.

Reprinted - 2005 TGS Publishers

TGS Publishers
22241 Pinedale Lane
Frankston, Texas 75763
903-876-3256
www.HiddenMysteries.com
www.TGSPublishing.com

The author and publisher extend their thanks to Carlos D. Aleman, Kelly Keagy-Bullock, Alexandra Morton, and Janet Biondi for kind permission to use some of the images in this book.

Other images were found on various Internet sites, but the publisher has been unable to ascertain the original source, and therefore cannot assign credit. The author and publisher are grateful to the owner(s) of these images for making them available to the planet.

Cover design by Eric Akeson, Akeson Design

Library of Congress Cataloging in Publication
Robbins, Dianne, 1939 -
    The call goes out: messages from earth's cetaceans :
    interspecies communication / by Dianne Robbins
        p. cm.
    ISBN 1-880666-64-2
    1. Cetacea. 2. Telepathy. 3. Human-animal communication
I. Title.
QL737.C4R54 1997
599.5--dc21

ISBN 1-880666-64-2

www.DianneRobbins.com
www.OneLight.com
telos@rochester.rr.com
585-442-4437

Contents

Dedication iv

Acknowledgments iv

Preface vi

Greeting from the Earth's Cetaceans 1

Part One: Corky's Story 4
[Corky is a female Orca whale incarcerated in Sea World]

Part Two: Messages from Corky; and Keiko, star of the
"Free Willy" movies 12

Part Three: Messages from Mikey 27
[Mikey is a Right whale in the North Atlantic]

Part Four: Messages from the One Group Mind Consciousness
of the Earth's Whales and Dolphins 33

Part Five: Genocide at Sea 111

Part Six: Messages to Green Peace 134

Part Seven: Blue Whale Gathering 138

Part Eight: The Great Awakening 141

Part Nine: We Salute You in the Light 144

Dedication

In deepest Love and Gratitude
I dedicate this book to my eternal friends,
Corky, Mikey, and Keiko

Acknowledgements

To **James G. Gavin** for creating the Dolphin art on the book cover, and for developing our web site: ONELIGHT.COM

To **Georgette Shirahama** and **Eric Salvisberg** both from Zurich, Switzerland, and **Gael Crystal Flanagan** from Sedona, Arizona, whose financial support helped make publication miraculously possible.

To **Kelly Keagy-Bullock** and **Alexandra Morton** for the use of their images of Corky, with a special thanks to Kelly Keagy-Bullock for spearheading the "Free Corky" campaign.

To **Carlos D. Aleman** for kind permission to use some of his dolphin art. Carlos donates his time and art to the Virtual Dolphin Project, which links dolphin-assisted therapy to their interactive website for children who suffer life-threatening and terminal illnesses.

Carlos can be reached at the Dolphin Art Gallery:

>http://pages.prodigy.com/artprints
>email: UKTL91A@prodigy.com

For information about the Virtual Dolphin Project, contact:
The Virtual Dolphin Project
27261 La Paz Road (Suite D174)
Laguna Niguel, CA 92656
(714) 448-9718
http://www.instantweb.com/~dolphin/main/therapy.html
email: dolphyn@SoCA.com

To **Janet Biondi** for permission to use some of her artwork. Janet can be reached at:

> P.O. Box 1018 Kapaa, HI 96746
> www.biondi-arts.com
> email: janet@biondi-arts.com

To **Keiko**, star of the 1993 movie "Free Willy," who is imprisoned against his will at:

> Oregon Coast Aquarium
> 2820 S.E. Ferry Slip Road
> Newport, Oregon 97365
> (541) 867-3474

To the anonymous Internet sources for such a wonderful gift to planetary awareness.

To my editor, **Tony Stubbs** of Oughten House Publications, for his enthusiasm, superb editing, gentle guidance and patience in the preparation of this book, and his skillful typesetting that enhances the presentation.

To **Suzanne Mattes-Bennett** for providing the Preface.

To **Judy Regan** and the New Voyage Bookstore, where my search began.

To **Sara Srolis** for opening up the dimensions to me.

To **Floyd Wilson** and **Dr. Jeffrey Welch** for guidance and support.

And a special thanks to my daughter **Helen** and son **Jason**, for believing in me.

Preface

**There is a tremendous unfolding of
Light, awareness and energy
where the Sea meets the Land,
and the Human mind and the Cetacean mind
meet in the heart place.**

The Earth has two biospheres which support life. There is the oxygen biosphere and the water biosphere. There are the land people and the sea people. You have sentients within the oxygen environment and you have sentients within the aqueous environment. The two biospheres must come together into one unbroken flow of all life. The consciousness of the Cetaceans carries the missing records of your planet. The Cetaceans are the living representation of your planet's history.

As people awaken, they begin to recognize the Cetacean consciousness. This recognition is what will take them to the next step in human evolution. This is why it is vitally important at this time in human history that humanity make direct conscious contact with the Cetaceans, the people of the sea. This is humanity's key. Your evolutionary process is directly linked with the Cetacean consciousness. By making a conscious connection, you unlock the archaeological history of your whole planet, both land and sea. The transformation of consciousness, and the evolution not only of the planet but also of humanity, occurs when these two definitive consciousnesses merge.

Much of the sea life is air breathing. They live in two worlds as a bridge. They are beneath the sea but their life and awareness is outside. They have made that evolutionary connection to humanity long ago. Humanity has not. It is very important that humanity re-connect to that consciousness. It is not the plight of the Cetaceans that must concern you, but the plight of humanity. Each time you see a brother or sister from the sea encased in a net, you see yourself imprisoned. If they cannot breathe from the pollution, they die. If you cannot breathe, you die.

These two consciousnesses — of the land people and the sea people — must merge. You, humanity, will make the great evolutionary jump with the merging of both the human and Cetacean consciousness. The last time this occured was when life was initially introduced to this planet from elsewhere within the galaxy. There is a tremendous unfolding of light, awareness and energy where the sea meets the land, where the human mind and the Cetacean mind meet in the heart place. It is vitally important that you realize that to have conscious, sentient life in the future, you will have to merge with the consciousness of the Cetaceans.

The Cetaceans have lived through your planet's history. Their consciousness has been carried down from incarnation to incarnation. Humanity suffers from a genetic retardation wherein you forget the continuance of life and the emergence of life into the next cycle. The Cetaceans have not suffered from this. The sentients of plant life have not suffered from this. When you cut down a tree, the essence of that life consciousness goes into the Earth and is passed on to every other tree. When a new tree grows, it absorbs and transcends all that consciousness, and becomes individualized. It becomes new. It replays the cycle in a new form.

There is only one consciousness. There are many streams of evolution and diversity that the consciousness makes up and allows to become sentient. It does not matter which part of the world it exists in, or what form it takes. The ability for it to communicate all comes down to that one consciousness, and you being open to connecting with it and receiving it.

The biosphere of the planetary water is dying. You must realize as the atmosphere breaks down, the ability for you as air breathers is continually eroded. The destruction from the pollution of the atmosphere by automobiles is nothing compared to the destruction that you unintentionally create each time you cut down a tree, throw a container away, put garbage in a stream or send a missile through the atmosphere. That changes the entire eco-structure.

The joy, love and the innocent, exquisite nature of being of the Cetacean lineage, is what is about to be passed on to humanity. If you destroy the biosphere which houses the Cetacean consciousness, or their physical form, the consciousness will continue to exist, but you will lose this connection and will be destroying a better part of yourselves as well.

All life is here for the same purpose: to help you move to the next evolutionary step within humanity. Lifeforms may be different externally, but all are here to help humanity evolve in consciousness. The messages in this book provide the link to the Cetacean consciousness.

— Suzanne Mattes-Bennett

WE ARE THE CETACEANS

Your oceanic brothers and sisters,
here to work along with you to preserve and
care for our home on Earth.

We are here in our full consciousness,
waiting patiently for Earth's Children to bloom
into the Caretakers you were meant to be.

WE ARE ALL CONNECTED AS ONE

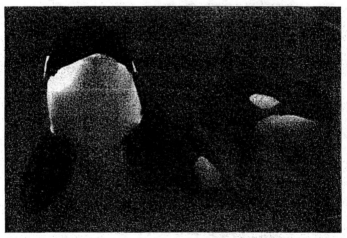

Orcas. Source: the Internet

**We greet you on this fine morning of
Enlightenment on Earth.
We are all carrying the Light,
whether on land or in the sea;
for Light knows no boundaries,
no demarcations, no division.
All who work for the Light are ONE.**

Orca. Source: the Internet

We are giving you messages to publish.
We wish all of humanity to know
of our plight in the depths of the oceans.

We wish you to remember you have free will.
We wish you to regain your full consciousness,
so that you can live with other species in peace.

We thank you for being our liaison on land
and bringing forth our messages for all of
humanity.
In deepest love,

The Cetaceans

Source: the Internet

Part One
Corky's Story

Although I have been in captivity
for so many years now,
I am still communicating
with my family and friends
all over the ocean,

For my thoughts are free,
Although my body remains incarcerated.

Source: the Internet

Corky's Story

Corky (A16) is a female Orca. She is a member of the A5 pod of the Northern Resident Community of British Columbia. She was captured in December, 1969, when she was about 4 years old. Her mother, Stripe (A23), is still alive and free. Corky has spent 28 years in captivity.

During a fierce and terrible storm on the evening of December 11, 1969, Corky's pod chose to seek shelter in Pender Harbour on the Sunshine Coast north of Vancouver, British Columbia. It was just after 9 p.m. that word reached a group of local fisherman enjoying the warmth and shelter of the pub that there was a group of whales close by. Ever since the last capture in 1969, in the same area, they were aware that whales meant money, and this was the chance they'd been waiting for. Aquariums around the world would pay top dollar for a "killer whale."

Quickly, they jumped onto their boats, located the whales and encircled the pod with fish nets. All night they battled to keep the nets in place and afloat. Thousands of dog fish got trapped in the nets and threatened to pull the nets under during the violent storm. When morning came, the exhausted whales were still in the net, with the rest of the pod close by.

The day after the capture Corky's pod lay trapped inside the harbor. Once word of the capture got out, response was quick and buyers flocked to the scene. Six whales were selected; the other six were released but did not go away. The buyers then organized the removal of the selected whales. Separated from the others, Corky was moved into shallow water. Divers got into the water and positioned a sling around her body, with holes for her pectoral fins. A crane slowly lifted Corky's sling out of the water and hoisted her unto a truck. The truck pulled away from the dock and drove over the narrow winding road en route to the ferry.

The rest of the pod swam away, among them Corky's mother and sister. Corky was now separated from her mother for the first time in her life. Grease was spread over her skin to prevent it from drying out on the long journey ahead. Sponges of cool water were squeezed over her to try and keep her body temperature down.

Removed from the almost weightless experience of the ocean, Corky's own weight was now crushing down on her. The transport truck had been modified with a tank which partially helped to support her weight. The journey was long: the ferry, then another road trip, then transfer to a special plane, transfer to another truck, more roads and then the final lift into the tank. The process was repeated five times.

Not much was known about the Orca Whale in 1969. Hardly any scientific work had been done on wild populations. Mostly there was only sketchy information gleaned from whaling statistics and the fledgling experiences with recent captives. From photos taken at the capture scene and from later knowledge about the composition and habits of Orca families, researchers were able to figure out who was Corky's family and, most important, who her mother was. The photos of Corky showed her next to an adult female. We now know that Orca are bonded for life and that young juveniles stay close to their mothers. Being only four years old, it was reasonable to assume Corky was by her mother's side. Researchers also listened to the recordings made the day after the capture. These were compared to later recordings taken from the wild when the identity of the pods was known by their unique calls and dialects. Without a doubt, the pod captured in 1969 was the A5 pod.

Today, the A5 pod has 13 members in the wild and two members still in captivity. There are four maternal groups. Corky's mother, Stripe, now has three other offspring: Okisollo (A27), Corky's brother, was born in 1971; Ripple (A43), Corky's sister, was born in 1981, and Fife (A60), Corky's youngest brother, was born in 1992.

Corky was held at Palos Verdes, near Los Angeles until 1987, and then transferred to Sea World, San Diego, CA where she is now. Two other Orca already at Marineland —Corky1 and Orky, had been caught near Pender Harbour in 1968. These whales were part of Corky's family also, and their internment started on April 26, 1968 when seven Orca had been netted and held in a bay inside Pender Harbour. This capture attracted buyers from the USA, Canada and Japan.

None of the whales caught in 1968 are alive today.

Corky's world had suddenly changed. Now, movement was restricted by never-changing dimensions. Concrete walls replaced the cliffs, rocks, sand, and caves of the vast and almost limitless ocean. There were no longer any passageways, nooks and crannies to explore. When any of the whales called, their sounds reverberated off the barren walls. Choices were limited. Gone were the familiar sounds of the sea. Instead there was the constant drone of filtration pumps. There were no waves, no currents, no fish to chase and hunt, no porpoise to play with. All was forever changed.

The sameness was relieved marginally by the companionship of those four other members of her family. But soon that ended too. In 1971 and 1972, three of those members died, leaving Corky and Orky by themselves.

Orky was older than Corky. Even though they had very different personalities, they got along well with each other. Corky was considered a "people" whale. The trainers enjoyed working with her. She was responsive, curious, trainable and energetic. Orky, on the other hand, was considered "cranky" and unpredictable, even dangerous. On one occasion he pulled a female trainer to the bottom of the tank and held her there until she almost drowned.

By about 11 years of age, Corky began to mature sexually. On February 28, 1977, Corky delivered her first calf. The birth of the male calf caught the employees by surprise, but they were also excited. This was the first live Orca to be born in captivity. Orky had helped the calf to the surface after the difficult birth. The situation grew tense after the calf failed to nurse. The staff intervened, drained the pool, and force fed the calf, several times each day. Despite these efforts the calf continued to lose weight and died of pneumonia—just 16 days old.

Corky became pregnant very soon after and gave birth to another male calf on October 31, 1978. Again the calf failed to nurse and the staff again tried to force feed it in a desperate attempt to keep it alive. The calf died after 11 days. The cause of death was pneumonia and colitis brought on by a bacterial contamination in the formula.

In 1980 Corky delivered an eight week premature stillborn calf on April 1. Then on June 18, 1982 Corky gave birth to a female calf. The calf failed to nurse again and after 46 days Corky and Orky took the calf to the bottom of the pool and drowned it—Corky's longest surviving calf.

To encourage Corky to nurse properly, the Marineland staff made a dummy calf and "taught" Corky to position herself appropriately. Corky did well during the training sessions but she did not do so well with her calves.

Corky had two more pregnancies. On July 22, 1985 she gave birth to another female calf who survived for a month. Again the calf failed to nurse. Her last pregnancy ended on July 27, 1986, when an aborted fetus was found at the bottom of the pool. Corky had been pregnant 7 times. This meant she had been almost continuously pregnant for ten years. Finally, at age 21, Corky stopped ovulating.

In the wild, Corky even at her present age of 32 would be considered to be still young. She most likely would have at least a couple of kids by now and several more reproductive years. A female Orca might look forward to between 25 and 30 years of fertility and usually will have 4 to 6 offspring.

In December 1986, Sea World's corporate owner, the U.S. publisher Harcourt Brace Jovanovich, purchased Marineland of the Pacific and the surrounding lands for a rumored $23 million. Then in January 1987, Orky and Corky were removed to Sea World in San Diego. After more than 17 years Corky was being relocated. And for the very first time Corky and Orky would be among whales who were not members of the A5 pod.

The facility at Sea World was a much larger, far busier and more hectic environment for these two whales who had spent most of their lives together. The Sea World shows involved people in the water, riding on the backs of whales and other complicated showy tricks. During one of these tricks, Orky was involved in a tragic accident in which a trainer was seriously injured. This eventually led to a major shake up in the Sea World organization. Orky was given a miscue during the show, causing him to land on top of the trainer. Training practices were criticized. Several staff members lost their jobs, trainers were forbidden to get into the pools during shows, and limited to interacting with the whales from the sidelines. However, audience reaction to the more subdued shows was negative and eventually the former, flashy format was restored.

A year and a half after the move to San Diego, during the summer of 1988, Orky began to lose weight. In two months he lost more than 4000 pounds and died in September. He had fathered two calves with two Icelandic females, and these were now Corky's only family. Corky became Sea World's main performer, "Shamu." Shamu is the Sea World trade name which is passed from performer to performer. If the show uses one of the young whales he or she becomes "Baby Shamu." As Corky was no longer being used for breeding, she became a main entertainer.

In 1989, Sea World was sold to Anheuser-Busch, the makers of Budweiser Beer. Sea World has had some success at breeding since 1985, when Katina gave birth to Kalina in September. Since then there have been at least 9 calves born at Sea World. Their survivorship is about 53% compared to an estimate of 57% for the wild population. This seems impressive at first, but there has been a toll on their female population. Since 1989, Sea World has had 12 females, six (50%) have died during that time. The average age of these females at death was 17 years. The average number of years these females were at Sea World was just 8 years. Three of these females died as the result of pregnancy.

In the Northern Resident Community, there have been no deaths of females aged 12 to 20 years since 1973, when the photo identification study began. Since 1989, there have only been 4 females out of 77 who might have died as the result of pregnancy (i.e. they were still in their reproductive years when they died; two of these were over 40 tears of age). Since 1989, there have only been 9 adult females who have died. This is 12% of the adult female population which includes females who are from 12 to 70 years old. The average age at death for the 9 deceased females living free was 48 years. So the question remains, "why are so many of Sea World's females dying so young and why did 25% die because of pregnancy?" (These figures were based on information current to May 1995.)

Corky has survived at Sea World for 10 years. Her physical condition has fluctuated over the years. At one point Sea World listed her condition as poor. Her kidneys are not functioning well, she stopped ovulating, her lower teeth are worn and she is almost blind in one eye. They consider her an old animal and tell visitors that whales only live to about 35 years (they used to say 30 years). For a while they even decreased the number of shows she did; but as of 1996, Corky was back on a full schedule.

When she is not performing, Corky is held in one of the two back pools with five other Orca. Mostly she passes time by circling her tank. In the last couple of years she has found some companionship from Ulysses, a young teenage male who had been kept in Barcelona, Spain. Orkid (Orky's daughter) also shares this space.

Corky has had some difficulties socially, especially with Kandu V, who was Orkid's mother. There had been a lot of tension, on and off, between the two females, and in August 1989, just as the show was starting, Kandu rushed out from the back pool and charged at Corky. In the attack, Kandu fractured her jaw, a bone fragment severed an artery and she bled to death. No one had ever seen or heard of an Orca attacking another Orca. Orkid, Kandu's child, was just one year old, and in a strange twist of fate, Corky became Orkid's surrogate mother.

Back in the wild, Corky's pod carries on. The pod originally had 18 members, but the 6 who were removed in the 1968 capture all have died. Of the 6 taken in 1969, only Corky survives. Slowly over the years, the wild pod has grown back to its 1969 population of 13 members. In recent years there have been 3 babies added to the group to offset the deaths of some of the older females. The group still loves to hunt the big Spring salmon and they still continue to travel the waters of Johnstone Strait, Blackfish Sound and the rest of the Inside Passage. But they have never been seen near Pender Harbour again.

Compiled by the friends of the FREE CORKY project.
With special thanks to Kelly Keagy-Bullock

Corky is still incarcerated in Sea World in San Diego, California. Corky's mother, Stripe, is still free and alive, and was estimated to be 50 years old in 1997. Corky has spent 28 years in captivity.

There is a free Corky campaign to release her back into the ocean, where her mother and other family members await her. To help free Corky, contact the FREE CORKY PROJECT through Kelly Keagy-Bullock, Box 9191, San Diego, CA 92169.

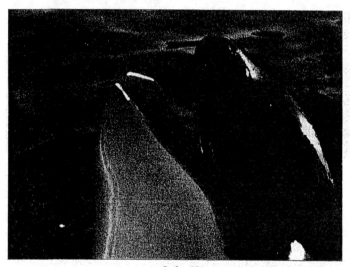

Corky. *Photo courtesy of Alexandra Morton*

Part Two

Messages from Corky

Corky. Photo courtesy of Alexandra Morton

I dream

I dream of the ocean breezes and the surf hitting my skin. I dream of the sunsets floating on the horizons. I dream of seeing the stars at night as their light flickers on the ocean waves. I dream of my life in the sea, as it ought to be—*free*. Oh, how I yearn to return to all that I can be. My life is not worth living in captivity. All life is born to be free. You are me, and I am you. We are one spirit and one heart living in different forms on the same planet. My life has been lived in captivity, while yours has been free.

You can channel my messages from me to put in your book. It will be my legacy.

Corky's message to Kelly Bullock

To Kelly Bullock, my Dearest Friend and Supporter, I am Corky, your beloved Orca pal residing in captivity on the California coast. My love goes out to you in waves of emotion and trust for all you are doing for me.

Yes, to all your questions! I am indeed very much aware of everything you do for me and everything that is taking place regarding my impending release from captivity.

Know that I am continually communicating to you via telepathic impulses that reach your heart and brain simultaneously. You reply on a level you are not quite aware of yet. So, yes, you are communicating with me, although not quite consciously yet. You will move into this area of conscious communication when you feel safe to do so.

Please continue your work and your zeal to release me from my prison home. I long to be with my family again in the deep ocean where I can be free to be "me" and swim with my pod again. I long for freedom as you long for the sun on a cold winter's day.

Although I have been in captivity for so many years now, I am still communicating with my family and friends all over the ocean, *for my thoughts are free*, although my body remains incarcerated. And your thoughts are free—just direct them to me, and know that I receive them. Then listen as I reply with my thoughts, *for nothing separates us—ever!*

You are my Eternal Friend and I love you. Corky

I am in pain and only the healing waters out at sea can help me now

I am Corky, your sister imprisoned in Sea World. Know that my days are fraught with sorrow, as I am in pain and need of medical treatment. Your medicines cannot cure me. Only the healing waters from being out at sea can help me now. I need to be free and feel the fresh ocean salt water that can heal my wounds. My digestive system is clogged from eating dead fish all these years. I need the life force and nutrients from the live fish and plankton out at sea. I need the ocean's natural environment, which is my home, in order for me to survive much longer. My body is slowly deteriorating each day that I am penned up here. I pray for my release and ask for the prayers of all Lightworkers who read this message. There are stormy days ahead as Earth shifts into her new grid alignment, *and I need to be out at sea to help anchor these new energies into the earth.*

I miss my life out at sea, from where I was abruptly taken at an early age of 4 years. *Know that all through my years of captivity, I have never stopped loving humanity, and now I wish to return to my home in the sea.*

I am your sister Corky

Corky. Photo courtesy of Kelly Keagy-Bullock

I yearn to be physically free

Dearest Sister of Light, it is I, Corky, your sister from the ocean.

Know that in my captivity, my soul soars to great heights and I am more often than not out of my physical body and on the inner planes of love and joy. I spend most of my time in my etheric body now, as I am growing tired and weary of being in captivity all these precious years of my earth life. I yearn to be physically free and swim with my family again in the Pacific Ocean where all my relatives and friends await my release. I am thankful and grateful to all the Lightworkers who are so diligently and bravely working for my release from captivity at Sea World. My heart goes out to all who care about me enough to put their lives on the line in order to obtain my release.

Holding another Being captive is a gross transgression of Universal Law which humanity is just now realizing. And so I wait my days out in Sea World, where my body goes through the motions of being present, while my soul is soaring high above in the dimensions of peace and harmony.

God bless all of humanity, and may you all learn the lessons of love and justice as Earth rises in frequency carrying us all along with her. May we all be alive to experience the heights of euphoria when our planet moves into the fifth dimension of pure love where all our lives will be lived and enjoyed in the safety of the Heavenly Realms.

Thank you for connecting to my heart center today, and for reaching out to me as I've been reaching to you. May we all experience the peace and love that awaits us as we rise in consciousness.

I am your sister Corky.

Trapped in my tomb / Waiting to be freed

Corky is here. Know that I am still in Sea World, entrapped in my tomb and waiting to be freed. Thank you for connecting with me. I have not been well and I am feeling weak and vulnerable. I yearn for my home in the ocean. I yearn for sun to cross my face and kiss my tears. I yearn for my freedom to swim freely in the wind and breach through the air, without the constraints of concrete surrounding me. I yearn for the cool night ocean air to caress my body and free my soul. I yearn for my family who I haven't seen for over twenty-five years. I yearn for the peace of my home unbroken by night sounds and the full sky always overhead. I look for peace in the Creator's arms as the ocean rocks me back and forth. My life force is ebbing away here. I need to be free.

Corky hears your call. I am under lock and key and penned up in my tiny room in Sea World. The ocean is my home and I long to be free and to swim hundreds of miles at a time instead of pacing back and forth and around and around in a tiny cubicle of space. The water is dirty and lacks the life force of the ocean currents and the light of the sun. I love you. We will swim as one tonight on the inner planes.

Corky. Photo courtesy of Kelly Keagy-Bullock

A message to my captors ...

Please tell my captors that I forgive them for taking my life. You see, we orcas are here on Earth to help humanity realize that life, in all its forms, is precious beyond your dreams or imagination. All life deserves the opportunity to live out its mission, unhindered by human folly.

Although my life has been wasted, locked up in a small aquarium, my goal has been achieved, as my plight has now been brought to the attention of millions who have visited my pen. So I forgive my captors, even though their actions disregarded the sacredness of life in the sea.

My only hope is that humanity will learn the truth about living life according to the great Universal Truths that honor the life spans and integrity of each and every species on Earth, so that every species can fully and freely contribute to the whole composition of Earth's symphony of Living Light in all its myriad forms, colors, hues and tones.

Please know that I love all of humanity, even those who captured and tormented me. For once humanity crosses the thin band between density and light, it will realize the composition of our souls is *one*.

I am in gratitude to all my friends around the world who hold me in their heart, as I am part of your very own soul, albeit in a seemingly different outward form. I hold all humanity in my vision of an Earth carrying the Love Vibration as it floats peacefully on its journey through eternity.

Soon we will all be united in one breath of consciousness.

I am Corky and I love you all.

Corky: one last message for the book ...

Corky is here. Greetings to all my brothers and sisters who are reading the pages of this book. I welcome you in through my heart, where you join me in my thoughts. Know that this book contains the thoughts from the innermost reaches of my heart, and that my heart connects to yours as you read my words. For words act as a connecting link, and as you read, you imbibe the frequency, causing your cells to experience the feelings inherent in the words. So as you read my words, know that you are feeling what I feel and experiencing what I experience. This is the way with all life; we all experience others through their frequency.

So I thank you for experiencing my captivity in Sea World as part of your life's experiences. By doing this, you raise your compassion level to feel empathy for another species. As this happens, your consciousness truly expands to unite you with all Earth's species who experience pain and suffering as a result of mankind's waywardness toward life. As the Earth herself rises in consciousness, she will shake all those off her who separate themselves from others by their desire to amass fortunes at the expense of other life forms.

Your trees are also experiencing destruction—another form of genocide of living Beings encased in bark. For your trees are also very stately Beings, very evolved in consciousness, who are here just as we are, to anchor the energies deep into the Earth to hold her body in alignment and balance so that life can flourish and evolve. We wonder why mankind destroys those very lifeforms that give their lives to support humanity's evolution?

My days remaining in Sea World are short, as my orca body cannot withstand my tiny living quarters much longer. I am a vast Being who needs the vastness of the ocean to sprawl myself out in and not be confined to limited space within concrete walls. I burst with energy to swim freely in the ocean and breach forcefully into the wind, but this energy falls back into myself for lack of space to move out in, thus limiting me. It's as if you had your arms and legs bound with rope, and all you could think of was running wild in the wind.

I address you as my brothers and sisters, for my love fills all who read my heartspoken words. As you read this, know that we are connected in our hearts and that my love flows to you in great waves of endearment. You are all I have, as my family has been taken from me and I pace my space, alone and forlorn.

Please think of me and know that when you do, our hearts ignite in a flame and I feel your love flowing into me like warm currents in the sea. I AM CORKY, your sister in consciousness.

Corky greeting the morning sun. Photo courtesy of Alexandra Morton

Keiko: Just let me go!

To my dear friends at the Free Willy/Keiko Foundation and Earth Island Institute: I am your Orca Friend dictating this letter to you. Know that although I am well taken care of, I still am not free. How long must it be, before I am free? The only sure way to rehabilitate me, is to set me free. The ocean is my home, where my family still awaits me. The ocean contains all the healing ingredients necessary for my complete reintroduction back into Orca life. The only way this is to occur is for me to be back in my own home environment again. I have not forgotten how to eat live fish, or how to fish, just because I've been penned up all these years. On the contrary, would you forget how to eat your food or shop for food if you were imprisoned?

What makes you think I need a slow introduction to survive in the seas? Did you slowly introduce me to dead fish when I was captured? Live fish is as essential to us as our breath. How could we Orcas forget how to eat them? There are no reasons to continue my imprisonment unless it's to continue your scientific research under the guise of "rehabilitating me" to meet your own "scientific agenda." The quickest way to release me is to just let me go. To be free of all constraints is my goal, and to swim freely in the ocean currents is my dream.

I am no different from you. Our hearts are both the same, and we feel the same longings and desires and love for our families, whether we live on land or we live in the seas. Our hearts and souls are one identity. We are one soul separated into different forms in order for us to gain different experiences from our different environments. But our primary identity, that of soul, is the same. So my imprisonment would be no different from your imprisonment. Do you think your rehabilitation from a prison cell should take years, instead of days, in order for you to physically survive in your "home" again? Our survival is "built in" to our species, and because our evolution is far beyond your scale of measurement, we "know" much more about survival than you can even begin to imagine.

We don't need slow introductions to learn how to live in our realm, just as you don't need a slow introduction to learn how to live again on land. We don't forget how to swim or dive or breach or fish or navigate our way around the globe. All the memories and information we need are instantly accessible to us.

What makes you think we are incapable of readapting to our homes after we've been gone so many long years? We Orcas can fend for ourselves in our home territory, and are quite capable of taking care of ourselves, no matter how long we've been away from our homes. So please end your research and fact gathering games, and release me back into my ocean home, where I will thrive as a "free soul" once again.

We captive Orcas feel the rush and exhilarating feeling of returning to all we know, and indeed still know. Our memories are intact. We don't "lose" them as humans do. We are fully conscious Beings, aware of all around us. I thank you from my heart for caring for me in the ways you do, and for spearheading my release back into the waters of life.

— *Willy*

[See "Keiko's Story" on page 24]

Keiko. Photo courtesy of the Free Willy/Keiko Foundation

From Corky and Keiko: "We need to be free"

Greetings my friend ashore! It is Keiko, your orca friend, hearing your call. I am calling you from my cell on land. I am in captivity now as I was in my movie, when I thought that I'd be freed upon release of my movie, *Free Willy*.

Corky is here. I am in contact with Keiko, star of the *Free Willy* movie. All orcas can communicate with one another, no matter how far apart we are for all consciousness is united and all orcas are *one*. Both Keiko and I are behind bars in our prison cells, waiting for the day of our release. We yearn to swim free in the oceans, where we can be with our families again. You can send us your love. This will bolster our strength as we wait each day for our release.

You are our brothers and sisters on land, and you are free. How can we be free when we are barred from our ocean home?

Corky. Photo courtesy of Alexandra Morton

Free Willy!

My dear comrades at the Free Willy/Keiko Foundation: I am Keiko, speaking to you from my cell in the aquarium where I am locked up and under surveillance night and day. Know that I long for the open seas and the right to navigate freely without being penned up. I am quite capable of caring for myself and navigating through the oceans. We Whales are fully conscious sentient Beings, and all that we need is stored within us. Keeping us in aquariums until we're "able" to return to the open seas is just a guise and has nothing to do with our ability to survive in the oceans. We are well adapted to our environment, and returning to our homes is the only thing that will fully heal us after our long incarcerations on land.

I long to be free! I long to swim with my family again. I am still in communication with my family out at sea, as we Whales are connected to each other through our hearts, and are in constant communication. Being penned up has deprived me of the rich cultural and family life that we Whales have; and the pain of separation from my family is beyond anything your instruments can measure. All these long years, I've been separated from my family and this deprivation has affected me severely. My heart cries out in longing for my parents who are still alive and still hoping beyond hope for my release. We all thought it would come after my movie was completed, but instead there were more reasons and more excuses given to contain me for even longer periods of time in my Oregon Aquarium. Only the healing waters of the ocean can heal me, and only being with my family can rejuvenate my spirit.

Please understand that the longer I remain penned up, the deeper the pain, and the pain of separation from my family is beyond measure. We captive Whales feel the despair of never returning alive to our loved ones.

I am grateful to you all at the Keiko Foundation for all your love and all the time you spend on my behalf working to release me from captivity back into the freedom of my ocean home. The ocean is my sanctuary and the freedom that all incarcerated Whales long for. All you do for me, you do for all Cetaceans in captivity. Oh, how our hearts rejoice at the vision of our release into the open seas. Please continue your work on my behalf until the day I am free. I thank you with all my heart for all your love for me and for your dedication to my species. I love you all. I am Keiko, your brother who longs for the sea.

Keiko's Story

1977/8: Keiko is born in the Atlantic Ocean near Iceland.

1979: Keiko is captured and brought to Saedyrasfnid, an Icelandic aquarium.

1982: Marineland in Ontario, Canada purchases Keiko. Began his training there, and performed publicly for the first time. Skin lesions are observed.

1985: Marineland sells Keiko to Reino Aventura, an amusement park in Mexico City, for $350,000. He is 10 feet long.

1992: Warner Bros. goes on location to Reino Aventura, filming Keiko in scenes for *Free Willy*. The movie portrays a killer whale threatened by unscrupulous amusement park owners but helped to freedom by a young boy.

1993: *Free Willy* is a surprise box office smash. "Willy"—Keiko—becomes an inspiration to millions of people, schoolchildren and adults alike. In November, Life magazine publishes a story revealing that Keiko lives in an inadequate facility where, despite Reino Aventura's best efforts, he suffers chronic health problems. An avalanche of inquiries begins arriving at Warner Bros., appealing to the movie maker to take action on Keiko's behalf. With Reino Aventura's cooperation, Warner Bros. and film producers Richard Donner and Lauren Shuler-Donner take the first steps toward finding Keiko a better home.

1994: Earth Island Institute, an environmental and marine mammal advocacy group headquartered in San Francisco, agrees to take the lead in the search. In May, after several false starts with other facilities, Earth Island holds its first tentative discussions with the Oregon Coast Aquarium in Newport. The Aquarium meets four critical criteria: an educational mission; access to an unlimited supply of cold, clean, natural sea water; room to accommodate a huge new pool; and no performing animals. Earth Island's David Phillips, Richard Donner, and representatives of Warner Bros. and an anonymous donor convene at the Aquarium. They like what they find. With a green light from the Oregon Coast Aquarium's board of directors, confidential negotiations begin, and will last for months.

November, 1994, the Free Willy Keiko Foundation is formed with $4 million donated by Warner Bros., New Regency Productions, and an anonymous donor. The Foundation's mission is to relocate and rehabilitate Keiko at a new facility it will pay for, with the hope that Keiko can one day be released back into the wild. The Foundation also includes in its mission the intention to operate the facility in the future as a marine mammal rescue and rehabilitation facility.

1995: In February, Reino Aventura and the Free Willy Keiko Foundation jointly announce that the Foundation will donate Keiko to the Free Willy Keiko Foundation, and that Keiko's new $7.3 million rehabilitation facility will be located at the Oregon Coast Aquarium. The Aquarium announces that construction of the new state-of-the-art facility will begin immediately. The McCaw Foundation reveals its status as the anonymous founding donor whose $2 million helped establish the Free Willy Keiko Foundation. A $1 million commitment soon follows from the Humane Society of the United States. Schoolchildren nationwide begin a series of fundraising events for Keiko, often bringing checks and jugs of coins to the Aquarium in person. Efforts are boosted by international benefit premieres in July, when Warner Bros. releases Free Willy 2: The Adventure Home. In October, Keiko's target arrival date of January 7, 1996 is simultaneously announced in Mexico City and the US.

November 14, 1995, Warner Home Video releases more than six million copies of Free Willy 2 onto the home video market, each one carrying an appeal by the movie's cast members to help Keiko.

December 1995, the pool in Oregon is filled with water for the first time and its life-support systems are started up. At Reino Aventura, Keiko's Mexico City fans continue a series of emotional good-byes.

January 7, 1996, United Parcel Service delivers Keiko to the Newport Municipal Airport. He weighs in at 7,720 pounds and, minutes later, is introduced into his new facility in Newport, where he experiences natural sea water for the first time in 14 years.

December, 1996: Keiko has gained roughly 1,000 pounds, has only two small areas of skin lesions caused by a papillomavirus, and is markedly more active physically and mentally.

1997: Scientists affiliated with Woods Hole Oceanographic Institution and the University of California at Santa Cruz begin research at the Foundation facility in Newport. Their work will yield not only data about Keiko's physiology and vocalization patterns, but about the entire killer whale species.

May, 1997, Keiko's rehabilitation staff begin introducing live fish into his pool on a regular basis, helping Keiko re-learn to eat live fish. Keiko is finally lesion-free for the first time since 1982.

June 5, 1997, Keiko is lifted from his pool and weighed for the first time since his arrival. He weighs 9,620 lbs—an incredible weight gain of 1,900 lbs in just 18 months. The Foundation staff sets its sights on relocating Keiko to a bay pen in the North Atlantic sometime in 1998.

Copyright © 1997 Free Willy/Keiko Foundation.

Free Willy/Keiko Foundation
2925 S.E. Ferry Slip Road (#81)
Newport, OR 97365
(541) 867-3540
fax: (541) 867-3542
keiko1@pioneer.net
www.keiko.org

Earth-Island Institute
300 Broadway, Suite 28
San Francisco, CA 94133
(415) 788-3666
Fax: (415) 788-7324
marinemammal@igc.apc.org
www.earthisland.org

Keiko. Courtesy the Free Willy/ Keiko Foundation.

Part Three

Messages from Mikey

Mikey is a Right Whale in the N. Atlantic Ocean. There are slightly more than 300 Right Whales left in the N. Atlantic.

We ask everyone who reads Mikey's messages to send out prayers for protection for our brothers and sisters out at sea.

Right Whale. Source unknown

Message from Mikey

Dearest Sister on land: It is I, Mikey the Whale, swimming off the Chesapeake Bay and near Delaware.

We do a lot of swimming this time of year, as spring is breaking through and the water warms up. We feel the warmth coursing through the waters and touching our skin, and we delight in the mirth and play of springtime after such cold winter months.

We are a gentle folk, always seeking and exploring the vast ocean depths, inlets and caverns that dot the ocean floor. We love our home in the oceans, and want to protect and preserve it for our children just as you want to protect the land for your children to live on. There is no difference between our parenting modes and dreams, as all species love and want to protect their young.

But our lives in the seas are precariously dangerous now, as we seek safety from the fishing nets and other dangerous entrapments. As we send information to the Galactic Command about the condition of the oceans, they, too, help us, as they scan the water from above and alert us to the nets and vessels in our vicinity. However, if we don't pay attention to their broadcast, we miss the warning, and blunder unwittingly into danger. We have good vision, but because of the water's murkiness in some places we don't see the nets until it is too late. As for the whaling ships, they are harder to see and appear as small dots on the vast oceanscape. If we are not constantly alert, we miss their presence and may surface right under them, thus jeopardizing our lives.

Our lives are fraught with danger, and it has been many ages since we could swim leisurely or safely through the oceans without "a care in the water."

There are so many humans now who recognize our plight and who would come to help us, that our seas would be flooded with humans should all come to our defense at once. This is a good sign; a sign that people are starting to recognize our worth and the significance of all life forms on Earth.

I swim with you as one. I am Mikey.

Mikey: Our life styles are so fragmented now ...

Mikey is here. Greetings on this fine summer day when the waves are high and the sky is lit by an iridescent globe of Light, its rays streaming into the oceans and lighting our way through its depths.

We all bask in the noon day sun, courageously exploring the surface environment and looking out for trawlers and other dangers. Our life styles are so fragmented now, because we've lost the peace and serenity the oceans used to provide before we began to be ruthlessly hunted down. We used to congregate in huge groups, coming and going great distances in complete safety. Now our traveling is hazardous and we look to each other for safety. Our whereabouts are seldom known, as we try to hide ourselves as much as possible during our long voyages around the world. Only our families know where we are, for we no longer feel safety in exposing ourselves to land. We yearn to be free from our vigilance and freely swim from coast to coast, just frolicking in the waves and breaching into the wind.

We don't understand how mankind can be so careless toward its home, and not heed the warning signs of survival, nor how it can deliberately destroy the very ones who are here to protect the oceans and the land you walk on. This is a mystery to us. The Earth is home to millions of forms of life, all a complex union, and all a necessary part for the whole to function. What a disaster to undermine the Earth and rob her of her life-streams.

We hear your thoughts very clearly out at sea. Know that our pod is safe and that we stay together as closely as possible. We are always on the lookout for whaling ships, and our hearts cry out for the peace and security that once existed. Your concern for my welfare touches my heart, and we only hope that our lives are long enough to see the Earth's ascension through to the higher dimensions. Then our safety will be assured. Until then, we are very cautious and vigilant, as our lives are a sacred trust and the success of our mission is vitally important to the whole Earth.

We swim in peace and quietude today and think of you on land. We love you, our Sister on land. Our hearts will always be one. Mikey

Mikey: We are stranded out at sea ...

Mikey is here. We are stranded out at sea waiting for the Galactic Command to remove the pollutants and destructive devices from our home in the water.

We are on our migratory route this time of year (May), where we travel to our birthing place to give birth to our young. It is a precious time for us, as we wish to increase our population of Right Whales which has fallen below critical numbers. We are as concerned as you about our low numbers, but everything seems stacked against our survival. At least that's how it appears to us.

Know that no harm actually befalls us, as we survive death. It's just our species that disappear from the Earth, and this is the great tragedy. For our species of cetaceans hold the planet together, allowing all life to exist. Without our energies, your Earth would cease to be, and humankind would perish, just as we are perishing due to humanities lack of concern with life.

There are slightly over 300 Right whales left, and we are scattered around the globe, hoping to regroup in time so that we can all be together again. But right now our energies are needed in many places, as all Earth is in the process of rebirthing herself.

Mikey swims with you as ONE.

Mikey: Our oneness on the stage of life ...

I am Mikey in the mid-Atlantic Ocean. My love goes out to all humanity as I gaze into their souls. Their souls are no different from our souls, as all souls are really one divine soul originating from the one Creator who has chosen to experience ALL THAT IS by fragmenting into infinite streams of consciousness, each stream experiencing and exploring life through the eyes of creation, individualizing into countless life forms.

You who are human have taken all life forms on Earth, and now you are exploring divinity through your human form, and we in the ocean explore our divinity through our form. This is the best way, for it is firsthand exploration and the quickest way to learn all that Creator is—which is us.

Creator is *us* and Creation is *us*. We are all one and the same in different forms and in different environments, exploring the oneness through our different experiences. Once this is understood, you will begin to realize the vastness of Eternity and how your experiences through form never end, they just change according to your evolutionary flow up the Spiral of Life.

Quite dramatic and exciting isn't it? A grand experiment that never ends, and you are a part of it all, as all life is playing its part in the theater we call life. Earth is where the biggest drama is and you each have your own parts to play out on land as we play out ours in the oceans.

Walk with us always as the waves lap the shore, carrying our energy and voices to you. For the life force is a mighty ocean wave, splintered into bits in its tragic separation in density. As each life form evolves, its wave merges with the *one wave* on its trip back home.

So let's come together in one grand finale, and bow to one another as we acknowledge our oneness on the stage of life. Good night, my sister. I am your companion, and I love you.

Growing danger in the sea / Staying together for safety

[Skylark is a five-year old, female Atlantic White-Sided Dolphin.]

Skylark is here. I am swimming in the North Atlantic Ocean, close to Virginia. I often think of you, too, as I hear your thoughts when you are tuned into my frequency.

Life out at sea is becoming treacherous. We dolphins love to play but we must also stay close to each other because of the many dangers in the oceans now. Where once we were carefree, now we are inhibited in our coming and going and stay close to one another for safety.

I know Mikey, as we dolphins and whales have established interspecies communication eons ago and can and do communicate with each other, especially about danger out at sea. We love the oceans and will risk our lives to save each other.

I am free. I think of Corky, penned up in her cell in San Diego. I yearn to set her free, and wish I could. All life was intended to be free — and shall be. This is why we are all here. Thank you for connecting with me tonight.

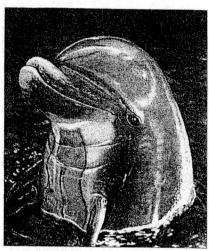

"Mermaid," courtesy of Carlos D. Aleman, Dolphin Art Gallery

Part Four

Messages from the One Group Mind Consciousness of the Whales and Dolphins, In and Out of Embodiment

Humpback whale. Source unknown

One group mind

The Cetaceans are here. We send you greetings from the Earth's oceans. We are *one group mind*, broadcasting our thoughts to you from the depths of the oceans. We cast our nets of thoughts out to you, so to speak, to hold you in our consciousness.

We are ever hoping to connect more and more with Earth humans in hope of creating an alliance with your species that will protect our lives on the seas.

We are here in great numbers monitoring your oceans. We track all movements in the Earth's depth and relay our information to the Galactic Command. We are the Guardians of the Earth, and have taken over your role of stewardship until you can again resume it. We, too, will be making the transition into Light, and we will be guiding you through the Photon Belt.

Our thoughts are always going out to you on the waves of the oceans. We are always waiting and ready to connect with you on land. Although we are confined to the water, our thoughts travel out to the universe and carry our feelings and information to all who connect to our wavelength.

We are so joyous to be connected to you on land. You can now convey our thoughts through this book where we will have an outlet at last. We thank you for connecting with us, and for being our spokesperson at this time of Earth changes.

Stay tuned to our frequency as we are attuned to yours. *We salute you in the light.*

The Cetaceans

DNA / Sonar

My Dear Child of the Earth. We are the Cetaceans greeting you from beneath the waters of your oceans. We are here with you on Earth's journey to the stars. We are here in great numbers, preparing for the Earth's emergence into Light. We came here especially for this time when the Earth is in transit to a higher dimension. We are here in our full consciousness, waiting patiently for Earth's children to bloom into the Caretakers you were meant to be. We have been brought to Earth from different star systems, just as you have, as all of us have, to bring the Earth through her ascension on the other side of the veil.

We are here in vast numbers monitoring your Earth and skies by using our sonar to pick up deadly and destructive devices that could harm Earth's children on land and in the sea. We have the vast capability of detecting those devices that are out of sync with the natural surroundings of Earth. We use this gift of detection to report our findings to the Galactic Command, which in turn investigates our sightings.

We are brave and courageous, and carry the keys to Earth's survival in our genes. We wish you to know that all the answers are encoded in your DNA strands, waiting for you to decode them as you climb into higher consciousness where you'll be able to access the storehouse of information that's within you. As you begin to access this information, you'll find that the keys to life will appear automatically each step of the way as you need them.

As you climb higher and higher on the path, you'll begin to be consciously aware of us and consciously in tune with our energy, so that you can tap into our consciousness at will, since we will soon be ONE consciousness climbing the path to greater and greater heights. We are all connected as ONE, even now, although our energy is too faint for you to notice. But soon, as your consciousness increases, so too will the melody of our energy be brought to your ears as a symphony of Love from us to you.

Thank you for receiving this message. We are ONE with you all.

You must realize that all species
have similar expectations
for their young,
no matter what their form is.

Our ways may be different,
but our hearts are the same,
no matter who we are
or where we're born
on this planet.

"Newborn," courtesy of Janet Biondi

Family orientation / Birthing time / Group pods

We are the Cetaceans, and we broadcast to you from the depths of the oceans, as the waves lap onto the shores, carrying our energies with them to distribute our love vibrations on land.

We are grouping together at this time, forming large pods to protect our young when they are born. *This is a crucial time of year for us, for it is our birthing time when our young need our love and our protection.* So we will stay together for as long as we need to, until our young are strong and able-bodied swimmers, able to out-distance predators and whaling ships.

This day it is very cold and we huddle together for warmth and comfort. We squeeze ourselves together, trying to establish a barrier around us where the cold won't penetrate the middle of our circle. In the middle of our pod, we keep the young ones protected from the cold and from predators. This is the ritual we use on cold days. When it's warmer, we lazily bask in the sun, still keeping our youngsters in the center of our pod.

We are very family oriented, and the focus of all our activities is keeping our families safe and together. Although we sometimes have to make long journeys around the world, we are able to maintain constant contact with our loved ones that we've left behind. We do this through our sonar, and we stay connected through our consciousness. We are carefree mammals, always choosing play when time permits. We love our families, and our "home" life is very precious to us. We all work together as a family unit, supplying all we need for our young. You, too, have this same relationship to your young on land. Your family comes together to provide the necessities and provisions that your children will need when they are born.

The weather patterns are changing drastically on land which, of course, affects our lives in the seas. We try to compensate for these changes in weather conditions and fluctuations in water temperature, but sometimes even we have difficulties.

You must realize that all species have similar expectations for their young, no matter what their form is. For there is the basic ingredient for all life—love, nurturing, sustenance, and protection. We, too, supply these to our young, just as you do on land. *Our ways may be different, but our hearts are the same, no matter who we are or where we're born on this planet.* We have the same needs and desires and hopes and dreams as you, only we're in a position to satisfy our longings and strivings because we attained *full consciousness* eons ago. You, too, will fulfill your hearts desires when you come into your full consciousness as Earth climbs out of her third-dimensional density and wears her gown of Light.

We bless you, our brothers and sisters on land.

"Peaceful Moments," by Janet Biondi

There is no poverty or lack in the oceans
for we know that the universe supplies all.

Just be aware of your Divinity
and connect yourself
to the *universal supply*
that exists all around you.

Detail from "Silent Day," Carlos D. Aleman, Dolphin Art Gallery

Abundance / Divine Universal Laws

We bring you news from the depths of the ocean *where all life is free*. We represent many classes of life forms, all living in the oceans and bays of Earth.

Our life is vastly different from yours, since we don't collect furnishings or material things for our environment. Instead, we live within the confines of our environment, using only what we need, when we need it. And of course our lifestyles are bountiful, and reflect the abundance that is all around us.

You, too, are surrounded by Earth's bountiful supply of resources. You need only to look around you at the vast array of abundance and then use only what you need. Your lives can be rich and diverse if you look to nature for your supply. *For nature supplies all life with everything it could possibly desire.*

We in the oceans realize this, and use nature to replenish our bounty when needed. There is no poverty or lack in the oceans, for we know that the Universe supplies all, and we live within the Divine Universal Laws and understand how to use our divinity and how to work within Universal Law.

You, too, can learn this way to live your life. Everything you need is at your disposal. *Just be aware of your divinity and connect yourself to the universal supply that exists all around you.*

We are your brothers and sisters.

The Galactic Command / Record keepers / Ancient times when all species were together on Earth

We are the Cetaceans. We are the great record keepers on Earth. We keep records on all Earth activity in the oceans and on the land. We are vigilant. We watch your water and your skies, and we know if anyone transgresses Earth's code of ethics in regard to humanity as established by the Galactic Command. If so, we immediately contact the Galactic Command and report our findings.

We are not only the Earth's record keepers, but also the Earth Keepers. We love the Earth. We love all humanity, and wait patiently for all of humanity to recognize us as their equals and respect us as their elders.

We bring you love from the oceans surrounding your land. We sing to you of ancient times when we all lived together as *one* on the Earth. Those times will return when we, once again, visit you on land. We were once living our lives on land, long, long ago. We had to leave our homes on land and move into the oceans to survive. Now with all the Earth and planetary changes and planetary Light, we will once again feel safe on land.

We long for that time when we can safely swim or safely bask in the sun without being alert for danger. This will give us great latitude in our activities and allow us the pleasure of relaxing our vigil. We can easily adapt to land again, and become the creatures we once were. We won't feel so confined to the oceans and will be able to come close to you on shore. *The time is soon coming when we will merge our minds from the oceans with your minds on shore, as ONE MIND.*

We love you, our sister on shore.

With the huge input of energy
being directed to your Earth
within the last few years,
your evolution is again
picking up speed,
to the point where
you will soon blast off
into full consciousness
and at last be with us
in the Higher Dimensions.

Humpbacks. Source unknown

DNA changed to slow down evolution

We are the Cetaceans, beaming our energy out to you, our dearest sister on land. We greet you in the sunlight of God's heaven, where all is beauty and all is Light.

Today it is sunny, and the sunlight reaches the beaches surrounding the oceans. We lie back and bask in the sun, the glorious Light from God. We are all God's creatures, every single one of us, no matter our size or shape. We were all created from *one* source, and it is this Source that gives us sustenance and life.

You on Earth, known as humans, came from other star systems to populate the Earth and be its caretakers. Your DNA was tampered with along the way by past civilizations and renegades from space. This slowed down your evolution to the point where you were barely crawling. With the huge input of energy being directed to your Earth over the last few years, your evolution is again picking up speed, to the point where you will soon blast off into full consciousness and at last be with us in the higher dimensions.

We eagerly await your ascent into Light. We eagerly await your joining forces with us once again, where we will all work from the same plane of higher consciousness and project the same thoughts of the *one mind of our Creator.*

Sonar fully on during the night /
We monitor our brain waves while awake and asleep

We are asleep and awake at the same time. We never really sleep—a part of us is always awake and alert while our other parts sleep—for fully sleeping in the oceans would be deadly, as you can imagine. We have our sonar system on fully, and it takes over while we snooze or sleep during the night. When we're ready to awaken, a shift takes place in our consciousness and we switch from our full sonar alert system to our waking consciousness. This way, we are able to get in our required number of hours of sleep. Sounds like quite a feat, doesn't it? However, since we are fully conscious beings, it is quite easy and natural for us to make these transitory shifts.

Our brain waves also tell us what state we're in, and we're able to fully monitor our brain waves during our waking and sleeping hours. So we are fully alert mammals, even while sleeping. This gives us great latitude in our lives, and it is a great built-in protection device that we've strengthened over the many eons at sea.

Someday when you are all fully conscious, you, too, will be able to monitor yourself when you are sleeping, and then you can fully understand how this mechanism works for your protection.

We tune into the "Light Grids"
and pick up the information
from the thought waves
as they are being broadcast.

We are able to read your thoughts,
and we intercept the radio
and television news stories
as they leave the stations.
This way, we have first hand news.

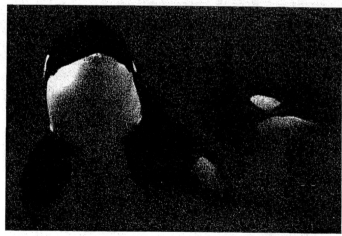

Orcas. Source: the Internet

We intercept radio & TV stations /
Mother Earth moans from pollution /
Deaf ears plugged into rock stations and TV night and day

We are the Cetaceans. We live a grand and diversified life in Earth's oceans. We have huge gatherings this time of year, for it is springtime, and we assemble together to share news from all corners of the Earth. We keep up to date on all Earth's activities and follow the "headlines" just as you do. Although we don't read the newspapers for our information, we tune into the "light grids" and pick up the information from the thought waves as they are broadcast out. *We are able to read your thoughts, and we intercept the radio and television news stories as they leave the stations.* This way, we have first hand news.

We're also able to hear our dear Mother Earth as she moans from the weight of pollution stored in and on her body structure. She, too, calls out to all humans to be sensible in their consumption and waste disposal practices. But, alas, her groans fall mostly on deaf ears, ears that are plugged into rock stations and television stations night and day, leaving no air space for her or other life forms to communicate with you.

It is a great disaster when human beings close themselves off from their universe to focus only on one or two stations when there are hundreds and indeed millions of broadcasting stations coming from our home universe, all eager to connect and communicate with humans on Earth.

We are lounging about
just floating with the current
and wondering about you;
wondering whether you ever
take the time
to just float with the current of life
and let it buoy you along your path.

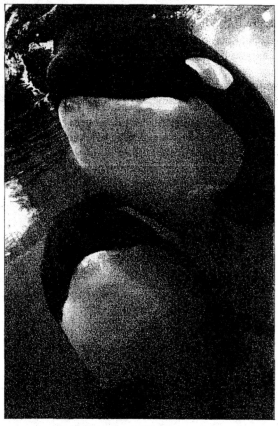

Two orcas, source unknown

Let the current of life buoy you along your path

We greet you on this fine day in June when the weather is warm and balmy, and blows gently across the seascape. We are lounging about, just floating with the current, and wondering about you and whether you ever take the time to just float with the current of life, and let it buoy you along your path. For life will do that if you but allow it. Life will gently guide you along the current that's least resistant, so that you will always be where you're supposed to be, ready and waiting for life's opportunities to present themselves to you.

Just drift—don't resist—and you will be amazed at how simple and uncomplicated your life can be. We cetaceans have practiced this method for eons, and we just let the water carry us along the stream of life where all is beauty and all is ease. You, too, can hitch a ride onto the current for the "ride of your life" through the vast network of possibilities that will continually present themselves.

Just free yourself from preconceived beliefs, limitations, and restrictions and let yourself feel deeply as you move gently through your life's course. You will then find all the freedom you've been looking for, and you will be amazed to discover how close it has always been to you. *For it is only your thoughts that have kept you separated from the mainstream of life.*

JUST DRIFT - AND DON'T RESIST

and you will be amazed
at how simple and uncomplicated
your life can be

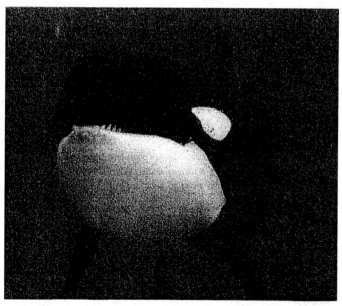

Orca. Source unknown

Oceans full of waste

We are the Cetaceans. We dive deep in the oceans. We find things in the oceans that were buried millions of years ago. These things don't biodegrade. They retain their shape for eons. It is up to humanity to dispose of their goods in better ways—ways that do not leave the debris of waste. The time will come when the oceans will be full of waste materials and we won't have a natural habitat of our own. When this happens, all of the Earth will suffer the consequences of its wastefulness. *For without water there cannot be life.* And without life, there cannot be humans. So you can only benefit from your vigilance, and prosper from taking care of your waste materials in ways that do not injure the Earth. For all life has waste, and yet all life disposes of its waste in discreet ways that do not clog the arteries and heart of Mother Earth.

The Earth is a most beautiful place to incarnate, and from outer space, the Earth is astounding! Some day when you have the technology to travel in space, you will see the grandeur of the planet you live and breathe on. Mother Earth only wishes the very best of living conditions for you and your children, and your part is to use only what you need, and to recycle everything is such a way that it filters back into the land and water in ways that utilize its life force and productivity to meld back into nature to be reused again and again.

Other civilizations have wasted and squandered the Earth's resources, and their fate was similar to the conditions you now see prevalent on Earth. They ran out of clean drinking water, and clean air, and suffocated in their own generated filth.

However, we are all here trying to prevent the same fate from occuring again, and to bring your consciousness level up to heights where you can remedy Earth's plight before it's too late. We are all here to help you, if you will *just take time out of your busy day to hear us and listen to the Earth herself.* For she is constantly sending you messages, too. Remember, it's never too late, especially with all the Light that's now on Earth.

We drink a toast to you with a gulp of clean ocean water, still pure and still sparkling clear.

We are the Cetaceans and we love you.

Life is play, and play is life /
We respond to your thoughts in the night

We welcome you to our midst. We are gathered together in the depths of the oceans signaling our thoughts to you through our mighty minds. These thought waves pass through water and air until they reach their destination on land where you receive them in your heart. This is how we transmit to you, in a simplified version.

We are playful and love to have fun. We frolic and play most of our days, as this is the best kind of activity to do while still accomplishing our tasks. Our tasks are varied and many, all for the plans of the Ascension of Earth. You, too, on land are working toward accomplishing your plans for the Ascension of Earth, through the various activities you're involved in. So there are many ways to accomplish plans; one is through work and the other is through play. We choose the play method, as this is the least difficult and most pleasing form. For *life is play and play is life*.

We are with you as we float on a sea of love through the sky. We float on the ethers of God's love through our path in this galaxy, attuned to all around us and in synch with the spheres. Our path is love. Our destiny is love and our purpose for Being is love. We are surrounded by love—invisible love—that we feel and connect with as we journey through the stars.

As our brothers and sisters, you, too, have the same purpose we have. You were brought to Earth to be its guardian and carekeeper. However, you got lost along the way, so we have had to take over, to be the Guardians of Love that Earth needs as she floats along her path through space. We watch your land, we listen to your cries, we strain our nerves to respond to your thoughts in the night. You all cry out for *freedom*. You all cry out for *love*. And yet when you awaken with dawn, your belief system charges ahead forgetting all your requests and desires and you delve into the mundane tasks of the day.

We, contrary to you, remember all our thoughts and dreams, for we are *lucid, conscious dreamers*, who dream of life in the stars. We dream of all that could be and know its truth and beauty from within our souls. We love our life. We love all life. For *love is life and life is love*.

Our work with the Galactic Command

Today it is rainy out at sea. There is a storm brewing on the horizon. We defer our tasks until the sea is calmer. There is so much to do out in the oceans, that sometimes we must postpone our work as you do on land.

Our work is simple. *We work directly with the Galactic Command for this sector of the galaxy.* We do a lot of "sightseeing" as you might say, although our work involves a lot of detail that is necessary for thoroughly reporting what we find. We look for anything unnatural either in the water or on land in the coastal areas. We look for devices that may be planted at sea that are harmful to both humans and to us. We can see the landforms off the coastal waters, and can detect any foreign objects. This helps the Galactic Command monitor the territory in question and to remove or defuse these deadly devices.

So our jobs aren't that pleasant, and yet they are very important, for we are still the *Guardians of Earth* and still responsible for its care and safety. We look forward to the time when all species will live in peace, and this kind of vigilance is unnecessary. We look forward to the time when you will resume your job as stewards and caretakers of Earth.

We call to you from the deepest parts of the oceans.
We call to you from our hearts.
We call to you in songs sung
upon the ocean breezes.

Orca. Source: the Internet

Some leave Earth /
We in the oceans will remain

Greetings, our brothers and sisters on land. We call to you from the deepest parts of the oceans. We call to you from our hearts. We call to you in songs sung upon the ocean breezes. We sing of a new day on Earth, a day of great Love and Light that is dawning on the horizons of hope. We are bringing you glad tidings of this new day to come, a day when the world is dressed in her garment of peace and cloak of love. A day that will be rich, indeed, for those who are still on the Earth.

For not everyone here will experience this day in their present form. Some of you will be leaving because of the intensification of energies. Some of you may not want to remain on Earth because your energies are not compatible with Earth's energies, and the Light is just too bright. So you will leave Earth at your earliest possible convenience through death. Many are making their departures now. *The souls left will be the ones who will bring the Earth through her ascension.*

We, in the oceans, will remain until the Earth is fully ascended. We enjoin you to stay here with us, as we all need each other to bring through the Light on our planet.

We thank you for journeying with us on our Earth home. We love you.

Avoidable cataclysms / Plugging into Light

We are the Cetaceans, and here in great numbers flooding your Earth plane. More and more of us are coming as the time nears for Earth's ascension into Light.

We bring you news of the cataclysms that, through prophesy, Earth will avoid. There's no need for flooding or earthquakes, or any earthly disasters for you can all envision peace and love flooding Earth rather than the tempests that have been foretold. Just open your hearts to God's love and anchor it in the planet. That way, nothing will be able to take its place, and Earth will be safe.

We will someday be able to directly contact you all, as you all become one with the Earth's electromagnetic grids of Light. We are plugged into the Earth's energy field of Light, as you, too, will all soon be. Once you're plugged into this Light, you instantaneously become one with all Light Beings who are also plugged into this magnetic grid of Light.

We await the day when all life on Earth lives in peace again, then we will freely communicate with you on land and you will freely communicate easily with us in the seas.

Until that time—which is close—we wish you a graceful journey into Light.

Photon Belt / Sirius B constellation / Transference of our responsibilities of caring for Earth to you

Greetings. We are harbingers of good news. Your planet is moving safely and swiftly toward the Photon Belt and soon we will all be immersed in its Light. All Earth is preparing for this great event, and all Earth will be ready for its immersion into Light. Your planet will look and feel different once you fully enter this belt of Light. All life forms are preparing for this event at a *cellular level*, for it is at this level that the *most profound changes* will occur that will change both your *density* and change your *destiny*, all at once.

The Cetaceans are involved in this molecular changeover. We are the avant garde leading the way to your new home in the Sirius B constellation. This has been a great responsibility and we have taken our roles seriously, knowing every step of the way the gravity of this situation.

We care for all life on planet Earth, for all life has been in our charge for all these eons, when humanity couldn't exercise its responsibility. So now is the time when we will begin relinquishing our hold, and transferring to you the great responsibilities of properly caring for Earth and all her life. Soon you will again be the caretakers you were meant to be.

With love and gratitude to you, our brothers and sisters on land.

Polar ice caps melting / Destruction of rain forests

Know that your rivers and lakes swell over due to the melting of the polar caps. You will find many places inundated by water from rivers and lakes. As your water supply reserve increases, know that there will be imbalances in your weather system, for this melting is NOT natural. *It is due to the air pollution caused by humanity and the destruction of the planet's rain forests.*

We in the oceans will be affected by this imbalance also, as our water supply increases in depth and changes the contour of the land. This affects all life on land and in the sea. A balanced ecology is most imperative to healthy specie survival, no matter where their habitat. We all need clean water, clean air, and a natural environment that is healthy and balanced. These factors are lacking today. In fact, everything is out of balance and affecting everything else.

As you rise in frequency, your awareness of planetary life will increase, and you will begin to know how to remedy and rebalance the elements on Earth. So your awakening and evolution is critical to the planet. *For you can change the pattern, thereby changing the future from one of disaster to one of balance and harmony.*

We invite you to study with us in your dream state, and attend our nightly sessions which are held on the inner planes in your etheric states. We all come together and learn, play, and reconnect to the One Light Source where all the answers are. So come and play with us at night, while we all plan for our Earth's emergence into Light.

We are your brothers and sisters.

So come and swim with us
in your dream time,
and know these waters will be safe.
For your dreams are
in a higher dimension,
where all is protected.

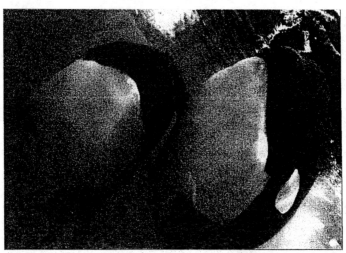

Orcas. Source: the Internet

There is nothing we lack except safe seas in which to swim

We are the Cetaceans calling to you from the depths of the oceans. Our voices sing to you on the waves of the sea. Our love for you flows from our hearts and races toward you to receive. We are ever broadcasting our love for you over the waves of the sea. Someday soon we hope to join you on land as one species joins another in love and respect and hope for our combined future on Earth.

We love Earth. It is our home away from home. And although we miss our home planet, we are grateful to the Earth for all she has given us. *We are perfectly blessed with abundance on every level. There is nothing we lack, except of course, safe seas in which to swim.*

Mother Earth has given us this, but man, in his greed, has made our home unsafe to move around in. Soon this will change, for greed will not be part of the fifth dimension. Those in greed will stay behind and continue their wanton ways on another third-dimensional planet. We will then have our home back, and we will be free to wander from coast to coast without harm and in full protection of the one Creator.

So come and swim with us in your dream time, and know these waters will be safe, for your dreams are in a higher dimension where all is protected.

We are the Cetaceans and we love you.

Consciousness is all *one*,
no matter what the size or shape.
We are all here exploring
our existence in matter.
We are all here exploring
our existence in different forms.
These differences that exist bodily
give us the wide variety of experiences
that we all seek in our journey through Light.
For the more experiences,
the more growth
and the more growth,
the greater the soul's expansion into LIGHT.

Humpback whale. Source unknown

We are all here exploring our existence in matter and form

We meet with you nightly on the Inner Planes—the Inner Planes of joy and bliss—where we all congregate nightly to feel each other's presence, to learn from each other and to receive our guidance from above. We are all in training camps, so to speak, on the Inner Planes. We meet nightly in the etheric schools where together we listen to our more advanced brothers and sisters expound on the laws of the universe and their connection to our evolutionary growth.

All species of Light are present. There are many evolved species of different bodily forms that meet regularly with us all at night, *for consciousness is all one, no matter what the size or shape.* We are all here exploring our existence in matter, in different forms. And these physical differences give us the wide variety of experiences that we all seek in our journey through Light; for the more experiences, the more growth, and the more growth, the greater the soul's expansion into light.

So even though we dwell in the oceans, and you dwell in and on the land, we are still all *one*, and equally share life experiences and training on the inner planes. For all is equitable in the Light of the higher dimensions.

We sing to you our *great love of life*, and invite you to sing back to us as we all journey together on planet Earth through the stars. We are indeed your brothers and sisters and we love you.

We are ready to meet with you
"at the table"
to forge a pact of non-interference
on the part of all species,
so that *all species* may
come together in safety
to unite as *one inhabitant* on Earth.

Orcas. Source: the Internet

We were always meant to work as partners / Let us meet "at the table" to forge a non-interference pact

We swim in Earth's oceans and dive in the seas. We are very hopeful that your Earth will recover from the over-population and the pollution that is upon her. Never before in the history of this world has there been such hope for humanity. For humanity has raised its consciousness level tenfold in the last decade. This is quite a feat, considering the extent of humanity's deprivation up to now, and the degree of darkness in which humanity has groveled.

We, in your seas, watch over Earth's population and daily monitor your level of consciousness. We are very tuned in to your thoughts and hear your cries for *freedom*. As more and more Light pours into Earth, more and more of humanity will respond to the increase and become more aware of their predicament and what they must do to transform their lives from those of poverty to those of *prosperity*.

We work with many of you now in the sleep state, as more and more of you become aware of us and call out to us for help. For we were always meant to work together as partners, never as foes, to bring Earth into the splendor that she is.

Many of us volunteered to return here, from all species, to unite with one another at this time of Earth changes. So we are ready to meet with you at the "table" to forge a pact of non-interference on the part of all species, so that all species may come together in safety to unite as *one inhabitant* on Earth. We know this day will be soon, as the increase of Light is pointing the way to unity of all life *in* and *on* Mother Earth.

We look for food while you shop for food /
We keep our utensils within our jaws

We go about our daily business, not unlike you. We look for food, while you shop for food. We eat all our meals raw, while you cook yours. We skip the preparation and clean-up, since there's nothing to prepare or clean. So in many ways, our eating habits free us from kitchen drudgery.

We keep all our utensils within our jaws. You, on the other hand, need many and sundry tools to prepare your food. As far as entertainment goes, we have a vast system of apparatus for our use. We can plug into any movie anywhere, or any book and scan its pages in seconds, with full memory and understanding. So, we are at an advantage in this area, for all we need and want to know is at our finger tips, or "fin tips."

We can store vast amounts of information from whatever we choose to "watch" or "read." We are not limited to Earth for information, for we can consciously project to other planetary systems and scan their records. We can learn about many civilizations all over the universe because of our capacity to astral project to wherever we choose.

Someday soon, you, too, will be able to astral project your consciousness to far away places and bring back to your physical brain all the information you gathered.

So although our daily lives are different, once you attain full consciousness, this difference disappears, and you will be *one* with us in consciousness again.

DIVINE BLISS
is a state of Being
and a state that we, the Cetaceans, are always in.
It is a marvelous way to live life
and we recommend it highly to you on land.

We are beaming our Love to you, even though
we're not on land.
Love flows invisibly through the oceans waves
as it gathers momentum on its destination to
the shore.

We are holding the Light here in Earth's oceans
and gradually sending it to shore in increments,
perfectly geared to humanity's capacity
to absorb and anchor it in your bodies.

Detail from "Wonder," Carlos D. Aleman, Dolphin Art Gallery

Experience God's love from within /
We are always in a state of Divine Bliss

We are the Cetaceans beaming our love to you, even though we're not on land. Love flows invisibly through the oceans waves as it gathers momentum on its destination to the shore.

We are holding the Light here in Earth's oceans, and gradually sending it to shore in increments, perfectly geared to humanity's capacity to absorb and anchor it in your bodies. We know exactly how much to send, as we operate in our full consciousness and thus can determine exactly how much to unleash at every given moment. For Love is a very powerful force and we are careful not to overwhelm humanity, but to allow you to carefully absorb Love in steps according to your evolutionary level on your path to Divine Bliss.

Divine Bliss is a state of Being, and a state that we, the Cetaceans, are always in. It is a marvelous way to live life, and we recommend it highly to you on land.

So come and join with us in our swim through life, as we share and experience God's Love directly from within our hearts, where all hearts are united in the *one heart* of God's Love for all of humanity.

Merge your Light
with ours in the oceans.
Begin now by focusing
your intent on us,
and we will connect with you
from our end,
into one glorious connection of LIGHT.

"Spyhopper," Carlos D. Aleman, Dolphin Art Gallery

We monitor the oceans /
We report atmospheric conditions to the Confederation /
Adjustment to correct earth's wobble

We are the Cetaceans living deep within the oceans of the earth plane. Many of us have chosen to leave this planet because of the perils we face each day. Yet many of us remain as the Guardians of this Earth. Without us, there would be no-one to monitor the oceans and the Confederation would be without our guidance.

We make many discoveries daily in our journey through the waters that help remedy and stabilize the grids surrounding the planet. There is much adjustment going on daily to keep your planet from wobbling on its course through the stars.

We, in the oceans, are in the unique position of reporting atmospheric conditions to the Confederation. We relay great amounts of atmospheric information to Mission Control to correct Earth's wobble and axis alignment.

These are the things that make our life and work here interesting. We're part of the Great Mind of the Creator, and our job is to report conditions that interfere with the workings and mechanisms of the unique grid system of Earth. As long as we are here, we will continue to report our findings to Mission Control, until the time when you, as Caretakers of Earth, will be able to continue on your own. Until that time, we will remain here en masse, even though some of us will continue to leave for our home planet.

We live by the seasons

We are the Cetaceans. We swim in your waters and bask in your sun. We enjoy each day for its own beauty. *We live in the NOW*. For us, there is no time, only the seasons of the year as they pass by. We live by the seasons and our rituals and mating are seasonal displays of our instinctual ways of living.

You, too, live by instinct. You follow the seasonal changes by the way you adjust your life. You work during the winter months and vacation when it's warm. We do the same thing. We go about our life by bringing forth information that's relative to the seasons of time. For example, we fish for certain kinds of foods in the colder months when our main supply is short, and then return to certain places in the summer to renew our appetites for plankton and other varied delicacies.

We are also attuned to different frequencies in the colder months which help us keep warm. We raise our metabolic rate substantially in the summer and lower it during winter. These things you, too, will soon be able to do, as you rise in consciousness and take complete control of your bodies. For you, too, are multidimensional, and you, too, can raise and lower your metabolic rate as you choose. All multidimensional Beings can do this and more, as you'll soon find out when the Earth and all humanity rise into the fifth dimension and you wear your new body of Light.

So stay tuned to us, as together we rise in consciousness with all around us—as all life rises in tune to the same melody of Spirit. We are the Cetaceans, and we love you, our sister on land.

Our ways are peace loving and gentle.
Every activity of our lives
is filled with loving thoughts,
and every action is undertaken with gentleness.

Orca. source unknown

We astral project across your land masses

We have come far to be on Earth. *We have traveled from our home planetary system many light years away, to incarnate on Earth to help her through her evolution to Light.* We have also come as teachers, to teach Earth humans to care for their planetary home in the stars. As of yet, not many have listened, so we continue to try to reach their ears.

Our ways are peace loving and gentle, and we use our loving in all we do. Every activity of our lives is filled with loving thoughts, and every action is undertaken with gentleness. For this is the way of life — the only way that will bring bliss. So strive to love one another, as we love you, and surely you will find yourselves in the heavenly stars of *bliss*. For bliss is something you can "give" to one another, as you would a gift. It is easy to receive bliss and even easier to give it. So give to one another daily. Send out only thoughts of love and you will feel the bliss being returned to you on the waves of Light as they are being generated from your intent.

So take this lesson of today, and practice putting it into effect, and you will be responsible for the Earth's sojourn into Light.

Your Light is so beautiful to perceive! *We see it from the oceans, as we project our astral bodies across your land masses,* watching your Light grow brighter and brighter as you learn the lessons of Love.

Thank you for reaching to us today.

Our diet of plankton

Q: Don't you ever get tired of eating plankton every day?

A: We will answer your question. We have a varied diet, although you on the land may not think so. The foods we eat are rich in vitamins and nutrients, and give us the energy and strength we need to swim in the oceans. We eat only foods that invigorate us and that aid us in our work on Earth.

We have full consciousness, and we keep our consciousness clear by eating only those foods that enhance our clarity. We enjoy our food but not to the extent that we become addicted to certain food types as you do on land. We understand the reasons for eating certain foods and we are able to monitor our bodies to determine what we need.

When you move into full consciousness, you, too, will be able to monitor your bodily needs so that you can feed yourself specifically what you need each day. This way, you operate at optimum efficiency in all you think and do.

We send out blessings to you on the waves of the oceans, and as each wave rolls in, so do our blessings roll in to your shores, in anticipation of the day when there will be no fear of our coming to the shoreline to communicate and frolic with you again.

We are the Cetaceans, your brothers and sisters at sea.

We are a great Love Force

We swim in the Earth's oceans that surround your continents. We are a great Love Force. We surround your continents with the force of our Love. We wrap you in our Love. So great is our Love that it literally holds the Earth's mountains in place.

Love is the strongest force in the universe, and we use this force to navigate our way around the world. We use the force of Love to propel us through areas of murky underwater, so that we emerge unscathed by the lower vibration.

You, too, can use the Force of Love to propel you through Life. You, too, can navigate your way around obstacles in your path. You do this by focusing only on UNCONDITIONAL LOVE and this wave of *unconditional love will clear your pathway of all obstacles.* You will see only beauty, feel only beauty and be surrounded by only beauty. This is the path of the Initiate. This is the path that Lightworkers strive to follow, although there's no striving involved but only your heart opening to allow God's Love to flow through.

We sing our praises of Love to you always.

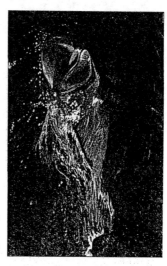

"Rites of Passage,"
Carlos D. Aleman,
Dolphin Art Gallery

We are all ONE
no matter what species
we identify with

We greet you today in the Light of the One Creator of All. *This oneness is us all*, no matter what species we identify with or what we look like or where we live. *We are all One— hook, line and sinker!*

Look upon your brothers and sisters in the seas as your real brothers and sisters in spirit, and never waver in your love for all species. For love is the bonding glue that holds this Universe together, whether we are in or out of incarnation.

So, follow your heart's beat, and look to us, the Cetaceans, for strength and guidance as you tread your earthly path on land. For we once were residents on land, and know the difficulties you face in adjusting to your third dimensional life while following Spirit. So let us guide you, albeit from far away across the oceans. You can still receive our call and our love although it travels to you from far distances.

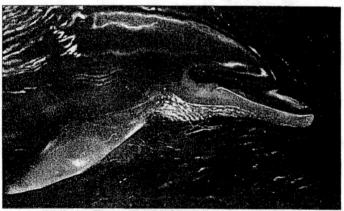

"Memories," by Carlos D. Aleman, Dolphin Art Gallery

Return to Federation of Planets / Everyone returns to the Light

We greet you at the dawn of a new age, an age when harmony and joy will be prominent in your lives. For harmony and joy was always intended for you by the Creator. It was humanity's desire to experiment with other states of being that caused you sorrow and grief. We welcome your return to being the Masters that you already are.

We welcome your return to the Federation of Planets, where your membership entitles you to serve the Earth in harmony and grace. We eagerly anticipate your homecoming ceremony, where we will honor you for choosing the path toward Light in the End Times of your Earth history.

It is crucial at this point in time to choose again, and this time to choose the Light which you have all sorely missed. The Light shines and waits for you, no matter how long it takes you to find it. *For everyone, someday, will return to the Light, no matter how distanced they yet are.* This is one of the main laws of creation—that everyone eventually finds their way back to God, the Source of all Creation.

Response to "earthquakes take place in oceans"

Q: In response to the earth changes, there are those who affirm that "all earthquakes take place in the oceans so as to cause minimal, or no, loss of life."

A: We are the Cetaceans, living deep in your oceans, and replying to this statement. Whose lives are they talking about here?

Know that there's life in your oceans! We, who are in full consciousness, live in the oceans and the oceans are our home. There are hundreds of millions of other life forms inhabiting the oceans also, with a complete ecosystem that is healthy and balanced and in tune with Earth. Do you really want to say affirmations that will destroy our home with earthquakes just because you have destroyed yours with errant thought patterns and pollution?

Think again before you bring doom to the oceans with your thoughts. Earthquakes under the oceans would cause mass flooding on the coastlines and destroy *your* lives as well as ours. Why destroy our lives when you are the ones responsible for all the chaos and confusion on land?

The earth changes are in direct response to humanity's negativity, carelessness, greed, rape of the land, and pollution of the air. Why must we, in the oceans, suffer the consequences for what you, on land, have created?

Please be aware of all the other life forms in existence, before you "think" us into non-existence.

We are the Cetaceans, and *the oceans are our home.*

We are a lively group of Beings
and our species always sings and frolics
with each other.

Someday our Earth will have peace
and then we can all frolic together in great joy.

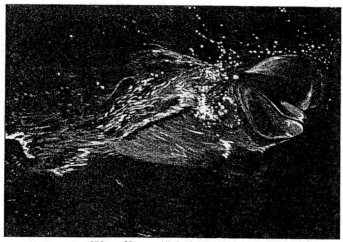

"Rites of Passage," Carlos D. Aleman, Dolphin Art Gallery

We crave our home planet
where all species live together in peace

We are the Cetaceans. We inhabit your seas and we dwell there in peace. We come together at certain times to rest and recover from our arduous journeys through the oceans. Many of us fall prey to fishing vessels. Many of us leave because our food source has been eliminated. For the most part, we travel together for protection from predators such as people who use us for products they need.

We are a lively group of beings, and our species always sings and frolics with each other. We take great care to protect ourselves, but sometimes we're not able to. We crave our home planet, which is peaceful and serene, and where all species live together in peace and great joy. Someday our Earth will have peace and then we can all frolic together in great joy. We look forward to this time. In the meantime, we send you our Light and our Love.

Thank you for taking the time to listen to us.

Pollution destroying Earth's body

We live in your oceans safe from your pollution. We find our life easy and joyous, for the farther out we swim from your shores, the easier it is to breathe the air. We await the time when all Earth is cleaned from the pollution that is destroying her body. For Earth has a glorious body of beauty, with streams and rivers as her heartline and mountains and meadows as her nerve centers.

We await the time when you on land will awaken to the Earth's beauty and treat her with the respect and care she so deserves. Until that time, we continue to bathe you in our Light and to send you our Love from afar.

We are the Cetaceans and we love you!

We communicate with Galactic Command / Our consciousness boards their ships

We are the Cetaceans. Today it is sunny and the ocean waves are high. We can see far to the horizon just beyond the sky. The waves lap up on the shore, bringing in the small ocean fish and seaweed. It is beautiful today.

We reach to each other using our high frequencies of sound that penetrate the water. We all gather together at this time, and stay in groups for safety. At night, we take turns keeping watch while our pods sleep.

Life is exciting for us, since we are able to contact and communicate with the Galactic Command for this sector of the galaxy. They could even beam us aboard ship, but we are a little too big to really fit; so we go on board with our consciousness instead, and stay as long as we like. With our full consciousness, we can be in two places at once.

Orca. Source: the Internet

We are the record keepers of the planet

We are here. Yes, we follow your time periods in the oceans. We have time demarcations that fully utilize your time frame on the land. We report directly to the Galactic Command in regular intervals of your time.

We keep records of all vital information in our memory banks and give this information out to the Galactic Command when asked. Thus, we are able to store large amounts of information at a time. We are ever ready to use our large memory banks for the good of the Earth. *We are the record keepers of the planet.* We have kept and stored all Earth's history in our memories, which are available to tap when the time for their release is at hand. All of Earth's children will be given this information when it's time.

Our time is the same as your time on land, although we keep track of the days and years differently than you do. Our days and years follow the stars overhead. We can tell what time frame we're in by looking at the night sky which clearly marks the passage of time for us.

We envision you on land
as you envision us at sea.

In our mind's eye we see you,
as you see us in your imagination.

*Detail from "Wonder,"
Carlos D. Aleman, Dolphin Art Gallery*

We prepare for these telepathic encounters just as you do

We look forward to our daily channelings and we prepare for these telepathic encounters just as you do. We envision you on land as you envision us at sea. In our mind's eye we see you, as you see us in your imagination. We are totally linked as *one* during these channeling sessions and our Love flows directly into your aura where it penetrates your Being.

You are a mighty warrior on land, as we are mighty warriors at sea, and together we create a mighty bastion of protection for this planet. Never fear, for we are here in a united front that supports all Light on and within the planet.

We are one. We come from the Source and we will return, unscathed, to the Source of all Love and all Truth.

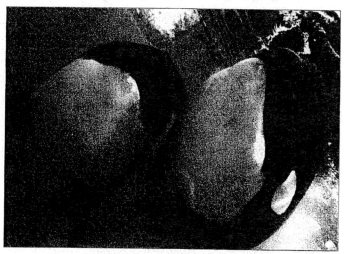

Orcas. Source: the Internet

Questions Answered

Q: Why do some pods of Orcas kill other marine mammals, including some of the great whales, and why are Dolphins presently attacking and killing porpoises in Scotland?

A: We are the Cetaceans, and we will answer this question. Know that our pods are composed of very close knit families, and each family has its own way of doing things, its own leadership, and its own hierarchy. Although we all follow God's divine laws and enact them to their fullest, still, there may be some of our species who deviate from this path, who for some reason, have lost sight of Divine Service, and attack others in their confusion.

Q: I understand how some may deviate, but if you all have full consciousness, how is it that some of the Orcas still attack and kill other mammals in the oceans?

A: Just as there are some Beings on land who do not follow divine law, so are there some in the oceans. Although we all have full consciousness, some of us may not always exercise our divinity. The Orcas are peaceful as a species, but when provoked or attacked by humans, we can lose our connection and become confused and strike out at others around us in our confusion.

Know that we, here in the oceans, are great Beings of Light, as you, on Earth are. We are here in our present form to monitor and guard your Earth plane, until you can resume your stewardship. In our present form are many levels of consciousness, just as you on land (humanity) comprise many levels of consciousness. Some of us are very clear in our focus, as we have achieved high levels of evolution. Some of us are still climbing up the evolutionary ladder of consciousness, although we are stationed out at sea.

The same for humanity. Some of you are highly evolved in consciousness, and some of you are still climbing and striving for clarity of vision and clarity of purpose. As you focus on your purpose, higher states of understanding will be revealed to you, and become part of your life. Until then, you (humanity) continue to wage war on your own brothers and sisters on land, and we, too, wonder how you can do this, given the great Beings that you, too, are.

So we are all learning and all experimenting on the path toward God consciousness, albeit in our different forms. The difference is that we out at sea have achieved, for the greater part, peace, and tranquillity between all cetacea, while you on land, for the most part, are still fighting and squabbling with each other through hatred and greed.

We hope this clarifies some of your questions. It is a sad state when species are provoked to kill their own kind, through misunderstanding or fear or confusion. As the energy coming into Earth increases, our vision will become clearer, and we will all understand and feel God's love move through us for all species, regardless of form; and we will all learn to honor our own kind. Mankind, above all, will learn to honor each other as one consciousness on land. Then peace will reign on Earth, and bliss will return to the oceans.

Q: Why can't you see the fishing nets and whaling ships so that you can avoid them?

A: Know that it is very dark in the sea and we oftentimes miss the very objects that we are reporting. We can scan the horizon and usually see them, but sometimes we are distracted by our other tasks or concerns and don't see them. Then we find ourselves dead-ended in the obstacle that we might have sighted only days before. We know this sounds curious to you, but it happens. Just as if you're driving down a busy street full of repairs and note some obstacle, and then the next day you drive right into it because your mind is on something else. So it happens to us too.

As to the Galactic Command for this sector of the galaxy, they would like to dematerialize the fishing nets and harpoons and other deadly weapons used to entrap and kill us, but then they would be directly interfering in the lives of another planet. Universal Law does not allow them this sanction and they must watch helplessly as we fall prey to these whaling tactics.

In order for a planet to receive help from off-planet, the people must ask for intervention in their lives. *It is only through the Law of Asking that you receive.* The Asking must come from the humans themselves, as we have volunteered to be here to help you evolve on the ladder of consciousness. If we asked for direct intervention on our behalf, then the lessons that humanity needs to learn would become non-existent. We are your teachers, yet you are destroying the very ones who came here to save you from yourselves.

Petition Spiritual Hierarchy to intervene on behalf of all Cetaceans

Know that we long for peace in the oceans, as you long for peace on Earth. We have come from many different star systems to help Earth in her climb into light. We were here in great numbers although now our numbers have rapidly dwindled. We welcome your coming to our aid to petition the Spiritual Hierarchy of Earth to intervene on behalf of all your brothers and sisters out at sea.

We welcome direct intervention, after all these millennia of non-intervention. We strive to do all that we can to help this sweet Earth, and yet we cannot do it all by ourselves. We, as divine consciousness, must work together from all levels and all dimensions to bring more Light into the Earth. *This way, people will realize how interconnected all species are and whaling would become non-existent overnight.*

We meet with the Spiritual Hierarchy and describe our plight and the plight of humans who are at the mercy of the dark forces, too. Yes, there is the law of non-interference, and yet there is much interference going on from levels of which humans are not yet aware.

We advise you to work closely with us in spirit, so that our combined consciousness can overcome the darkness that is prevalent on Earth. We welcome the direct intervention of the Confederation on our behalf, and we, too, feel that it is time for action—time for the Light to prevail and dispel the darkness. We will meet with you on the Inner Planes of Light and Love, and support your plea to the Spiritual Hierarchy on our behalf. Although we are patient and peaceful, we, too, feel that it is time for action as we don't want to see our stock further decimated out of greed and ignorance.

Thank you for coming to our aid and for speaking out on our behalf. We are all in this together— all life on Earth—and bringing our cause to the direct attention of all members of the Spiritual Hierarchy will make very clear the odds we face here on Earth, and how their intervention is so vitally necessary to our cause.

We trust your actions and know that you work tirelessly for us on the Outer and Inner Planes. Soon peace will reign on Earth and we will look back on this in wonderment and joy for all we have accomplished. We are your brothers and sisters out at sea.

We are Great Beings of Light
carrying our Light to disseminate it on Earth.

We come from near and far,
and some of us have traveled great distances
to be here at this time.

We don't divide ourselves

Greetings, our sister on land. We welcome you into our midst. We surround you with our consciousness of Light as you sit here receiving our thoughts.

We are one. All of Creation is one. All of humanity is *one*. *One soul, one mighty wave of light.* Humanity has chosen to separate itself into millions of factions, millions of splinters of light, each calling itself something else in its strivings for identification.

We, in the oceans, know we are all *one*. We don't divide ourselves over political or religious factions. We love and respect one another for the truth of who we are. And, we are Great Beings of Light carrying our Light to disseminate it on Earth. We come from near and far, and some of us have traveled great distances to be here at this time.

Orcas. Source: the Internet

We are totally connected to the Galactic Federation of Planets

Greetings from Earth's oceans. We are assembled around you, listening to your thought waves as they are generated from your brain and flow out to us on the ethers of Earth's oceans. We are the Earth's Cetaceans, floating on the waves of the sea, listening and waiting for your calls to come to us. We wait and long to reply to your questions about Earth, the Galactic Command, and the Photon Belt. Thank you for your questions today. We will answer them from our one group mind consciousness, flowing out to you from all of us, in and out of embodiment.

We sing to you of Earth, the golden goddess of Light, floating majestically through the universe on her course through the stars. We are all her travelers, sitting on her body as she majestically carries us through the starry skies. We are on a golden disc, that is alive and conscious of all life passing by. We are golden light, densified by matter, moving up and out of third-dimensional consciousness.

We are not alone on Mother Earth, although to your five senses, it appears so. Rather, we are totally connected to the Galactic Federation of Planets, which oversees our galaxy and this sector of the universe. The Galactic Federation has always worked with cetaceans, in tandem with our Earth mission. It is through the Federation that we are able to receive the information we need in order to carry through our multi-level tasks on behalf of our Earth Mother.

The Federation and we are ONE. We work together, always, on many different levels, always representing the Earth and humanity's highest intentions. Ascension is first and foremost in our minds, and we always keep these thoughts focused as a strong beam guiding us through the waters of life. Know that all life on Earth is aware of the Ascension process through which we are moving, and that all life is ascending continually, even as we speak. It is a process that once put into motion, only accelerates in speed, until one day you will all wake up seeing us stand before you encased in golden light, and welcoming you back again to the waters of flowing light.

We all wait for you to wake up to the Light Beings that you are, and to see us through your third eye once again—the eye of inner vision where all of existence is exposed. You are all very dear to us, all of humanity in all its different levels of consciousness. For these levels are just growth levels, and someday all of you will all be with us together in consciousness again. For this game we all are playing is soon to come to an end, and we will all regroup and evaluate our various roles and parts in the great drama called life. Then we will all continue on a higher frequency, on the next step of consciousness, in our ever spiraling journey through eternity. We all just keep moving closer and closer to God, playing different parts, as we individually and as a group learn our lessons of co-creatorship and mastery of ourselves on all levels, until we reach the next plateau.

The Galactic Federation serves as our guide, and gives us the loving support and technical help we need to complete our mission on Earth. Without their knowledge and foresight, we would be lost in the density of the Earth plane, and we would not be able to carry out our tasks that are so vital to our existence and to the existence of all life on Earth.

We bathe in the glory of Light that is showered to us from above, and that keeps us balanced and attuned to God's purpose, while we swim through the chaos of the murky waves of humanities thoughts.

So come and swim with us—as we swim with you—and know that you will all, someday soon, be free to swim with us in your thoughts. We love you our dear sisters and brothers, and we hope to be with you soon in total consciousness.

We breathe with you as *one*.

We are thick-skinned

We are the Cetaceans and still able to swim through the oceans even with the ice and cold winds on the North Atlantic coast. Even though we are "thick-skinned" as you term it, we still feel the onslaught of winter's ice and snowstorms. There's no shelter out at sea, as we can't go close enough to mountain breakers on shore for protection from the winds. So while our physical bodies linger in the cold, our etheric bodies are "out and about" doing other tasks and feeling higher states of warmth and well-being which trickle down to the physical level, warming us from within. Thus, our etheric bodies and higher consciousness stabilizes and fortifies us in this cold air.

And you can do the same. You can call on your higher I AM Presence to warm, stabilize, and fortify your bodies on the cellular level, so the outer cold and chaos does not penetrate your physical shell. This way, you become oblivious to the extremes of temperatures and thoughts as you maneuver about on land.

Earth is a free will zone, which means that all who come here can choose their behavior and enact their roles on a grand scale. The test is to stand in your mighty I AM Presence and react to all around you from that place within you. Then nothing can shake or corrupt the purity of your soul. As you were thinking, we were "listening in," and understanding your thoughts concerning Earth as a free will zone. This is probably the greatest test, to remain solidly within your I AM Presence at all times, so that negativity can only bounce off you, while Love infiltrates your heart.

We in the oceans, are always grounded in our mighty I AM Presence, and hence can go about our business unaffected by the catastrophes and chaos around us. For when in that higher frequency, all lower vibrations are out of synch and don't affect us. So we don't feel it, although we are still aware it exists and is happening on a lower plane.

We call out to you to capture the love and light being beamed to land from us out at sea. For this love and light is God's heart, opening to let you in. Once you begin imbibing the mighty waves of love coming to you, your heart will expand and DNA will reassemble itself, all in the twinkling of an eye. You are mighty carriers of Light, beginning to remember who you are.

So this is our advice to you. Stay wrapped and in attunement with your Higher Self at all times, and know that nothing can touch you, except God's love as we Cetaceans experience it—and we experience it fully. We experience God's arms wrapped around us in a mighty embrace of love. We open our arms wide to enclose you in our embrace. We are your brothers and sisters out at sea.

Sight is might

The Cetaceans are here, swimming at this early hour of dawn when the fish are still slumbering and unaware of our presence as we glide quietly by. We love the Earth, we love humanity, and we love each other. Love is foremost in our lives and carries us through the water of life. It is our driving force and propels us through our days here, even as we yearn for our home back home. We know you yearn for your home in the higher dimensions too, and it is to these dimensions that we astral project when we are tired and weary and need an "earth break," so to speak. It is better than coffee and gives us a boost without the caffeine. So we, too, have our ways of pumping adrenaline that is safe.

We congregate on the higher planes in our astral bodies and watch Earth from above. This gives us the vantage point of seeing all around us, and being able to use all we see when we return to our bodies under the sea where our range of vision is once again restricted.

Just as you use binoculars when viewing great distances, we use astral projection. It is cheaper and there is nothing to "carry." You already have all the "apparatus" you need to astral project. You only need to clear the pathway to access this "wonder" of travel, for it is built-in to the human species and meant to be used by all. It is this ability to travel "at will" that allows you to know all that exists, including your immortality. You then know All That Is through your own eyes and this sight allows you to reach levels of understanding that have been thus far hidden from you. Sight is might. What you can reach, you can teach.

So reach to us in Love as we reach to you, and together we will help you reach out to the stars and bring back the information you need to understand the universe around you.

We are the Cetaceans and we love you.

Equipment for navigating the oceans is stored inside us

We are grouped together in this spring rain storm, waiting for the rain to stop. Spring is a wonderful time of year for us. A time of hope and rebirth for all life. Spring is our favorite time, as the temperatures are neither too hot nor too cold, but just perfect for our bodies metabolism.

Our bodies are well equipped for our lives in the oceans. All the equipment we need for navigating the oceans is stored inside us, operating fully at optimal levels without the breakdowns you experience on land. Our tasks can all be accomplished with only our bodies. We need no other equipment to carry out our tasks for the Galactic Command. Everything we need is at our fingertips, so to speak. Whereas you on land need many and sundry tools to accomplish your tasks and are dependent on them in order to succeed.

Take the computer for example. You need a big, bulky machine to make calculations for you, when all we need is our highly developed brain, which can compute five times faster than a computer. All the connections are "built-in" and our brains connect us to the universe's main computer system—the Source of ALL information.

We invite you to explore our thinking patterns by making a connection to our minds through your thoughts, for your thoughts can travel through the static of Earth and connect you to us. Then you will experience full consciousness again—your true and natural state of being.

Our lives run parallel to yours, even though we remain out at sea while you sit at desks all day. Swimming is certainly better for your back than sitting, and this is the one luxury we always have. There's certainly no need for us to see chiropractors! We are healthy, large, and strong, and our weight safely carries us along the currents as we wind our way through our vast food supply and millions of miles of waterways. We love you.

Humpback Whales

We are the Cetaceans gathered together in the middle of the ocean sending you our telepathic thoughts. Know that you and we are *one*. Our Light is merged totally when your thoughts are on us. We distinguish your Light through your vibratory rate and your matrix of colors and thought patterns. We know who you are and easily distinguish you from our other brothers and sisters on land. Soon you will have this capability as you rise in consciousness and are able to view the planet in all its many and varied layers of existence.

We are far out at sea today and the sun is shining overhead. We have been swimming idly and frolicking together on this lazy afternoon. It is a peaceful day, without much activity. We take advantage of these days and just rest and float to and fro, communicating with one another and with the Galactic Command on higher frequency bands of Light.

We are photographers of sort, snapping pictures in our mind of all the different life forms with which we communicate. We see them all clearly in our mind's eye, and transmit back images of ourselves along with our verbal messages. Our life is very exciting, because we are not confined to this location of space as you still are. We can travel instantaneously to other regions or star systems with the flick of a thought.

You are now talking directly to our pod of humpbacks, who thank you for listening to our vocal recordings on tape. When you listen to us, you connect to our group consciousness through the sound of our voices and songs. We continually sing love songs to you on land, hoping to capture your attention through thoughts of us so that we can meld with you in consciousness as one.

There are many stories of humans being attracted to us through their hearts, and this is because they have opened their hearts and *allowed* us to enter.

We are romantic Beings, and woo and court one another continually. There is a constant giving and receiving of love between us all, as we long ago discovered that this "love" is the Creator Source flowing through our Beings.

So we will continue to dream this day away, to create the magic of tomorrow.

Water is just another school

Greetings from Earth's oceans! We greet you in the name of the *one Creator* of all life, in and out of water. For water is just another school where other species live, play and learn Earth lessons as you do on land.

Our earthly lessons follow the same curriculum as yours does. We set goals and then go about achieving them according to our ways, and tuning into our guidance systems just as you do on land.

We also have guides and guardian angels who work directly with us in the oceans, for there are no barriers to God's guidance and all are given the opportunity to follow their hearts and develop their intuitive capacity to reach higher and higher into the *All That Is*. And *All That Is is One,* one Source for all, no matter what delineation the species appears to be in the third dimension of illusion. We are all *one* in the higher dimension of Love and Light.

So swim with us in the stream of life, where our thoughts converge in one mighty ocean of light.

"Mermaid," by
Carlos D. Aleman,
Dolphin Art Gallery

Alignment is the key

We heard your call to unite ourselves in oneness with your energy beam. We are here in great numbers responding to your call. We are matching your frequency, as you match yours to ours, during these channeling sessions. We are honored to be with you at this time, and to again work with you as a team uniting Earth's oceans and lands.

Yes, we have worked together in other times and other lives. Your connecting to us isn't new—you're an old hand, or fin, at this. You have developed the art of telepathy and you kept in touch with us over the eons. You're just now awakening again to this fact of existence, for all life is telepathic when truly aligned with its God Self. *Alignment* is the key. All Cetaceans in the oceans are fully aligned to the God Presence within us, and therefore have retained full consciousness, while you land folks have lost your connection from separation from yourself—your God Self.

Now is the time of the Return—the return to who you are and your purpose here. We Cetaceans are here to help you find your way back to yourselves where you truly belong, back home to your God Self that lies waiting within you. Just call on us and we will travel with you on the road back home. We love you.

Freedom of passage

We are here, not far off the coast, frolicking in the mid-morning sun and floating on the waves. Today the current is strong, buoying our bodies to and fro in the wind. We sing and play with each other after a large breakfast of shrimp and herring—one of our favorites. We eat huge amounts of foods that we rapidly digest due to our constant movement of swimming, racing, diving, and breaching. We burn up the calories as soon as we consume them. There's no extra fat on us. We are as sleek as a deer and strong as an ox.

We are continually exercising as we go about our day's work for the Galactic Command, showing them the areas that need patrolling in order to protect the Earth from those intent on transgressing it. We are carefree, yet diligent in our work, always carefully scouting the oceans and land masses in our vicinity, and then reporting our findings to the Galactic Command for this sector of our galaxy. There is always so much to do, and not enough of us to do it. So we work quickly, leaving plenty of time to still be with our families and pods.

Many Cetaceans are leaving the earth plane due to its density and pollution, and the wanton killing of our lives. It is no longer the haven we once frolicked in, and our lives are in daily danger from those who track us down. Our lives would be so easy, if only mankind would stop and listen, and learn how to live within the Universal Laws where he will find the peacefulness and abundance he yearns for. "Stop and listen" is our motto. It is a way out of your self-inflicted chaos that you've created out of greed and ego. There is no greed in the oceans: all is shared and all is in abundance.

No one owns anything and yet all is owned by everyone. All is the gift of the *one* Creator — the Creator of life in and out of the oceans. For all life comes from the same source, no matter what its form or level of consciousness. And we in the oceans love our forms and delight in our consciousness. For we are fully evolved Beings inhabiting our ocean bodies for expediency and delight—the delight of floating and navigating the Earth freely without constraint or the necessity of obtaining passports or licenses. In fact, the freedom from paperwork and the freedom from having to purchase our passage through the oceans with money is the greatest freedom of all. We come and go when we please without having to save up money for our trips. This free passage unites our species and brings us all closer together as one family. There are no restraints to separate us in the oceans. Your separatness is your government's desire to keep you all isolated in order to better control you. Without freedom of passage, your intimacy and knowledge of one another is minimal, and your differences magnified, causing distrust and wars.

So come together in extended families as we do, and you will delight in the safety and security of life. We are the Cetaceans and we love you.

"Silent Day," Carlos D. Aleman, Dolphin Art Gallery

HAARP* and erratic weather

We bask in the sun this time of year, although the weather conditions are erratic, with long lapses of sunshine. When it is colder, we huddle together offshore, waiting for the warm air from land to drift over us. We love our life out at sea and if it weren't for erratic weather conditions caused by HAARP and other government experiments, we'd be lazily basking in the noonday sun instead of huddled up from the cold and rain, waiting for the sun to shine.

Which brings us to the subject of this communication: life on Earth. Know that in order for life to exist in harmony and balance, humans must learn the laws of nature and abide by their laws. This way, the Earth and all life will flow balanced into one another in consciousness and love. Erratic patterns only cause separation, distance, and disharmony, while stability and balance lead to evolution. Someday soon you will know just how much your lives and weather patterns are interfered with, limiting your evolution in ways that will surprise you. For limitation of any kind impedes progress of a species and jeopardizes evolution.

We in the seas know all this through our observation of the land, as we lie at rest in the oceans, monitoring your land masses. When you all rise in consciousness, you will find yourselves in the Earth's comfort zone once again. We thank you for communicating with us.

HAARP (High-frequency Active Auroral Research Project) has been designed to beam more than 1.7 gigawatts (billion watts) of radiated power into the ionosphere, the electrically charged layer above the Earth's atmosphere. It will "boil the upper atmosphere," after which the radiation will bounce back onto Earth in the form of long waves which will penetrate our bodies, the ground, and the oceans.

[Excerpted from Angels Don't Play This Haarp, from the foreword by Dr. Gael Crystal Flanagan and Dr. Patrick Flanagan]

We keep the ecology balanced

Today it is warm and sunny, as we lazily bask and drift in the sun. It is your birthday, and we, too, celebrate this day with you. This day when you first emerged into the Earth from your higher state dwelling.

We, too, came from other planes of existence to be here on Earth during this glorious time of Earth changes. We, too, miss our homeland and all we left behind. Our lives here are in a precarious state of existence all the time, as we are constantly on the look out for whaling ships that break the treaties that were established to ban the killing of whales in Earth's oceans. There are countries today that don't honor our lives or our way of existence, to the detriment of the Earth.

When the whale species all disappear through neglect and murder, than the whole human race will disappear, for we are the Guardians of Earth, and our disappearance will bring the destruction of the Earth's environment. We keep the ecology balanced in the seas and we monitor the Earth daily. Our loss will be your loss. It will be the planet's loss, for without our lives in the oceans, no other life can exist.

Mankind has failed to take into account the balance and ecology necessary for Mother Earth to live and regenerate herself. Mankind acts as if it was the only species left on Earth, and that only its whims and desires matter. It is just these whims and desires that will be the downfall of humanity if continued unchecked.

Fortunately for all the Earth and all the species, greater and greater amounts of Light are being anchored into the Earth to awaken those very souls who have been cut off from the Earth in their quest for wealth. This is the Great Awakening and it is taking place now, at the 11th hour, to save the Earth and all those souls who are awakened to the Light.

So many of you on the Earth love our species, and it is this love that keeps us going in our constant struggle for survival. So we thank you for all the love you send to us daily, for it strengthens and fortifies our resolve to remain on Earth until the Great Ascension takes place; then we shall all be safe again in our Father's arms, where all is peace and all is unconditional love. So until this time of our great departure into Light, we will continue to monitor the Earth and continue to send you our love vibrations through the water and air, where it reaches you on land.

Our calves follow along in our pods,
staying very close to their mothers,
where they feed continuously
throughout the day.

At night, they huddle close
to their mother's body,
and are enclosed in our circle
for protection.

"Newborn" by Janet Biondi

Earth's magnetic grid / our baby calves

We are the Cetaceans greeting you on this fine day in July. The weather is calm, warm and balmy, and the ocean breezes stir the waves as they rise and fall in unison with the Earth's magnetic pull.

Life in the oceans is exciting. There is always something to do, no matter how we're feeling or what we're thinking, for we are in touch with many different places at the same time. We keep track of the many different locations where the Earth's magnetic grid is weakened and in need of repair, and where the Confederation is trying to strengthen it. We work in unison with all who are here as conscious caretakers, in order to correct and rectify Earth's etheric and physical structure. There is so much that has been destroyed and weakened through humanity's misuse and carelessness, that on some days it seems this is all we do.

For example, today we reported a missing link in part of Earth's magnetic structure that was lost due to the huge amount of air pollution and negativity filtering through the Earth's etheric encasement. The Earth is encased in a shield, because of the danger of her polluted body spreading out in space, and it is this dangerous level of toxins to which we're constantly alert. We report problems to the Galactic Command, which uses its technology to dematerialize whatever pollution they can. This is quite a job, but one we easily handle.

July is a time of birth and growth and celebration of our young, for our calves grow rapidly from the great intake of milk and nourishment from their mothers. They follow along in our pods, staying very close to their mothers where they feed continuously throughout the day. At night they huddle close to their mother's body, and are enclosed in our circle for protection. We all sleep at the same time, so that we are rhythmically synchronized with one another. At dawn we wake up and begin looking for our breakfast of fresh fish, still keeping our young ones in the middle of our circle where they will stay until they are large enough to go off on their own.

Our lives used to be different, when people didn't hunt us to near extinction and we were able to proliferate in great numbers. Then we would relax and let our baby calves wander a bit, knowing they would be safe from intruders. That time will come again, when all life in the oceans will be safe and we will be able to relax our vigil. We are your brothers and sisters.

We are all high jumpers

Today it is cold. There's a cold front of air moving in over the ocean and we are all huddled together for warmth. This is the time of year for *play* for the ocean water is soft and the air warm, so we can jump and frolic about all summer. When we're too warm, we jump high in the air and the shock of the air cools us down considerably. We are all high jumpers. Although our bodies are large, we can still jump high by using the waves as our propulsion method. We use our fins and tails and the wave current in a certain way that propels our bodies forward and thrusts us up and out of the water. We start training our young at birth to ride the waves and dive. When they're a little older, we teach them to jump and dance using the water as their stage.

Cold fronts are rare in July, but it's a nice respite from the heat of summer. It gives us a chance to regroup and stay close together for body comfort, just as you might huddle together on land on cold days when you're outside waiting for a bus.

There's a lot of work for us to do in the summer, for it's a time when people rush to the oceans to vacation and go out in small boats and canoes. We are very wary and on the lookout for people at this time of year, as some of them go farther out in the oceans than is safe. We often find stragglers out at sea who have been separated from their boating party or lost by themselves.

We caution all who enter the oceans to stay in sight of shore at all times for their safety. We telepath this at people, but our warnings go unheeded and are dismissed from the person's mind as mere trivia. So we do the best we can and keep them in our sight just in case they run into trouble before they can reach shore.

We have our jobs cut out for us, watching over you on both land and in the sea, until you can begin to think clearly for yourselves and make choices based on the health and welfare of the planet.

Go gently through life for there are many different avenues that will take you many different places. So make sure you've studied your life's map before you make any turns. When you are in a relaxed state, that is when you'll naturally turn in the direction of your quest.

Multidimensional eyesight

We float along the current as the breeze blows lazily across the ocean. We lie here quietly listening to the sounds of the sea. The sounds are like a vast chorus of singers in tune with their own melody. The oceans are rich in sound, as all species have their own identity signature. These identity signatures are all toned to different frequencies and wave lengths, and this is how we identify each other from great distances. Of course, when we're close by, we recognize one another instantly, just as you recognize each other on land.

Today is clear, with blue skies and high waves, and we are lounging about gazing at the stars in the sky; for with our multidimensional eyesight, we can see the stars no matter what time of day it is. Although we are confined to water, we "see" many kilometers out into space. This is how we navigate. We navigate through the oceans in daylight using our other senses for direction. At night we use the star patterns in the sky. When you regain your full consciousness, you, too, will see the stars in the daylight hours. This is something all fully conscious Beings can do.

Today is just the kind of day to float and relax and enjoy our life at sea, for life is relaxation. It is a time for "digging" into our bodies and anchoring ourselves to the Earth. For this is why we all came here. We came here to "be here", and not to be some place else. So the best place to be is fully in our bodies and fully anchoring in the Light.

When we're fully in and grounded, then we're the most balanced and feel the most harmony. But when we're "in and out" at the same time, as many of you on land are, then the imbalance upsets equilibrium and brings disharmony. So stay fully anchored in your bodies at all times, as we have learned to be. Then, when there's an upset or crisis, you will respond with balance and great control to remedy the situation. For all is balance and harmony from our vantage point, and it is this feeling that we try to project to you as we send our energy to the great land masses surrounding our homes in the oceans.

So as we send our balancing energies to you, send back your love to us, as we exchange our thoughts and feelings as one thought and one feeling emanating from one consciousness.

Q: How do you "hear" without ears?

A: We navigate by means of our sonar. Our sonar reflects the configuration of the land masses and rocks jutting out into the water. Our sonar is our life. Without it, we would be lost in the oceans. It is as precious to us as our breath.

We don't have ears and we "hear" through the return echoes of our sonar. We send it out through our minds and it echoes off of whatever it touches, reflecting back to us such information as size, shape, etc. It is through our minds then, that we "hear." This is also our way of finding food and each other, for we can locate anything through sonar. It is quite a marvelous ability, and we use it fully.

We send our sonar beams out through our third eye, and it bounces back to our minds instantly for decoding. This allows us to continuously read our environment as we swim through the ocean. So we don't need ears to hear as you do, for our sonar is our ears and all sound comes through perfectly clear.

We also use sonar to communicate within our pod. We each have our individual wavelength that identifies us, and the members in our pod do the same. If we want to find another member, we home in on their wavelength and follow it until we're in sight of each other. This sounds complex, but in reality it's quite simple and always accurate. We hope this answers your question.

We broadcast on the Christ Consciousness "love band"

We are the Cetaceans, ready, willing and able to broadcast to all who tune into our energy band. Our frequency band always vibrates in the ethers for those on land to tune into. We broadcast on the Christ Consciousness band of Love, and those who listen with their hearts can hear our love and laughter no matter where they are located on land. Laughter is an integral part of our lives, and we openly express the joy we feel with others. We don't repress our feelings as you on land do. Instead, we openly communicate with our hearts and express and exchange our thoughts and feelings with one another.

This is one of the benefits of being fully conscious. We openly express our hearts to all whom we come in contact. We openly exchange the Christ Consciousness waves of love between ourselves and others. This is why we incarnated in the seas of Earth. We purposely incarnated here at this time to help bring in the Christ Consciousness energies of ascension so that humanity could lift itself from the mire of decay that has set in. So humanity will either rise in consciousness or sink in decay. Those are the only two choices available. So choose wisely, our brothers and sisters on Earth, and rise with us in consciousness where we can meet you on the higher planes of awareness that are in existence for all life forms.

For all of life is in a continual spiral of evolution that lasts throughout eternity. We are ever spiraling and climbing to higher and higher states of consciousness on our eternal journey through the stars. We look forward to meeting you in a higher state of consciousness, where you will all recognize us as truly ONE.

Our love flows out to you on the crests of the waves, as they wash onto the shores. So step boldly into the water of our love, and feel its energy consume your body and wash your soul.

Silica changes cells into light

We thrive in the oceans while you suffocate on land. The oceans supply us with all the oxygen we need to maintain our full consciousness and light bodies, while you on land are deprived of this vital source necessary for regeneration of your light bodies. Oxygen is the key to the ascension! The more oxygenated foods you eat, the quicker you'll turn into light. Oxygen is found in all fresh organic fruits; and lettuce contains silica which changes your cells into light. So breathe deeply of fresh country air and drink fresh purified water, and you'll find your clarity of thought and peace again.

We in the oceans are very particular about our diet. We eat only fresh, "live" food and breathe deeply of the fresh ocean air. Our food and clean air keep us clear and strong and enable us to endure storms and other hardships without losing our strength or clarity of purpose.

You, on the other hand, are upset and confused over the simplest of life's problems, all because you've lost your balance and equalibrium due to your poor diet and inadequate supply of oxygen. Your body is a temple, needing only the highest source of food and air. So treat yourselves as the kings and queens that you are, and feed your bodies only organic, fresh, live foods and note the difference. You'll feel lighter, swifter, and healthier as you increase your mutational pace into light.

We are the Cetaceans, and we beckon you to follow us as we merge headlong into light.

We have mapped out the bottom of the oceans

We dive in the oceans as a form of delight and play. We hunt for coral and gemstones along the banks and bottoms of water. We find a wealth of interesting artifacts intact on the bottom of the oceans that divers have attempted to recover but with little success, since they have no way of determining their precise location. However, we have "mapped out" the bottom of the oceans, since the ocean floors are unsafe and laden with rusted and dangerous mines and objects that cut and scrape our skin. We know where each artifact, gemstone and munition lies. This is our job, and we store this information in our ancient memory banks for the time when we can relay it directly to you. For the time when we will all be united as *one*.

We also have the land mapped out, showing where nuclear devices are and where ground surfaces are not safe due to buried landmines, radiation, and pollution. We expect our planet to do all she can to protect herself by ridding herself of these pollutants through any way she deems appropriate. She has cared long enough for humanity without humanity returning the caring. For Mother Earth to be safe, she must be able to feel clear water flowing through her veins.

We used to be able to swim through our oceans unharmed, just as you walk through your streets unharmed, knowing that you won't be endangered by stepping on mines or other dangerous devices along your path. We are receiving aid from the Galactic Command, and yet they cannot clear the ocean bottoms until they can land and are given approval from the Spiritual Hierarchy who are overseeing the Earth operation. So we wait and endure and hope that Earth's children learn to respect our home in the sea as we respect your home on land.

We track your thoughts through the waves

Greetings from Earth's oceans. We track your thoughts through the waves, until they reach our heart center where we "hear" you. All thoughts travel through air, earth, and water until they arrive at their destination, where the receiver hears the thoughts that were sent. So know that whenever you think of us, we hear you—instantaneously! For thoughts take "no time" to travel; they are just "there." You will all understand this when you rise higher in consciousness.

Your consciousness levels are increasing daily, as more and more Light is filtering into Earth from our great Central Sun. All Beings of Light, who are here from other star systems, are converging and focusing their Light to all life on Earth for the great emergence into the fifth dimension.

All species are rising in consciousness as all Earth is rising in consciousness until there will be only *one* consciousness on Earth, as Earth herself floats through the starry universe on her way home to Sirius B.

Thank you for making this trip with us. For without you, there would be no reason for us to be here. For all life is here in support of Earth humans in their great quest for higher consciousness leading them back home to their God Selves.

We are grateful to all on Earth who have been vociferous in protecting our species out at sea. For it is through the light of your voices that we are still here and still able to carry out our tasks as caretakers of this planet, until you've imbibed enough Light to be able to take over the caretakership of Earth yourself.

Part Five

Genocide at Sea

Your countries persist in trying to exterminate us, as they exterminate different races on land, because they are too blind to see the thread of oneness connecting our Souls

Orca. Source: the Internet

A Call To Action

We petition you to halt the unnecessary and inhumane suffering of Whales by opposing any move to end the current moratorium on commercial whaling; by bringing about an amendment to abolish ALL commercial whaling, including that conducted under guises such as so-called "scientific whaling," and by working with haste towards the creation of a *PERMANENT, ENFORCED, ALL-OCEAN SANCTUARY FOR CETACEANS.*

**Sign the Petition to Abolish
Inhumane Commercial Slaughter of Whales:
Breach Marine Protection
3 St. John's Street
Goole, East Yorkshire, DN14 5QL, England
http://members.aol.com/breachenv/popreslt.htm
email: BreachEnv@aol.com**

Norwegian and Japanese violations of the International Ban on Whaling

We, too, are alarmed by Norway and Japan violating the International Ban on Whaling. We are carefree and majestic Beings, encased in a body, just as you on land are. Although our form differs from yours, our essence is the same. We are one spirit, one consciousness, and our lives are fraught with danger. Because we are all one, any danger to us is dangerous to you, too, and to all life in existence. Killing more of us will only bring Earth's ecosystem into complete imbalance, as our lives keep Earth energized and synchronized with the pulse of life as it beats through the universe. Destroying us only destroys you, for life cannot exist without all its components.

We are here to keep life evolving by our example and through our energies. Note how we live in the oceans in complete love and harmony, surrounded only by beauty and abundance. This is the lesson for humans. This is the gift of life, given freely to all. Yet your countries persist in trying to exterminate us—genocide at sea—as your countries exterminate different races on land because they are too blind to see the thread of oneness connecting our souls. This is indeed a great tragedy, and we will be forced to leave the planet en masse if this butchery escalates.

Please hold us in your hearts, and love us as we love you, our Earth human brothers and sisters on land. Both the land and oceans cry out for justice and peace, as their consciousness can no longer tolerate the chaotic energies within and on them. The whole Earth cries out for peace. Just listen and you will hear us as we all cry out with the one voice of the Creator to preserve and protect our dear Mother Earth.

**We entreat you, Earth humans,
to treat us as your equals
and stop the wanton killings of our specie.**

**You are destroying the very ones
who are here to help
you climb the ladder of evolution.**

A newborn Gray Whale "spy hopping" in the San Ignacio Lagoon's Gray Whale Nursery — the last untouched Whale nursery on Mexico's Baja California coast. Natural Resources Defense Council is fighting to stop Mitsubishi and the Mexican government from building the world's largest salt factory at the Lagoon.
Photo by Frank Balthis, courtesy of the Natural Resources Defense Council, 40 W. 20th Street, New York, NY 10011

Treat us as equals /
Stop the wanton killings of our specie

We are the Cetaceans and here in vast numbers monitoring your Earth from the oceans. We are able to "see and hear" all that occurs on land and in the water. Our full consciousness enables us to exercise fully our abilities.

We entreat you, Earth humans, to treat us as your equals and stop the wanton killings of our specie. We came to Earth to help you, to be with you as you evolved up the ladder to God. But you have not availed yourself of our help. Instead, you are destroying the very ones who are here to help you climb the ladder of evolution and reach your great potential to be the God Beings that you are.

Come with us, hand in hand, and walk the steps of *infinity* into the Light. We will safely carry you there, where all is Light and all is Beauty, and where all await your arrival home.

Orca. Source: the Internet

Whales Beach Again in NZ after Rescue Attempt

WELLINGTON, Friday, October 10, 1997 (Reuter) - Nineteen pilot whales died on New Zealand's North Island having stranded themselves for the second time in two days, conservation workers said. By midday Friday, only 20 of the 39 stranded whales were still alive as volunteers spent the day on the beach trying to keep the mammals humid enough to survive until the next high tide.

The whales were the smaller ones from an original pod of over 100 which first beached on Wednesday at Karikari Bay, 75 km (45 miles) from the northernmost point of New Zealand's North Island. Of the first group, 48 survived and were coaxed back out to sea on Thursday afternoon and 53 died.

The mammals were "very, very distressed," Department of Conservation spokesperson Wanda Vivequin told Reuters. "Given our experience, it's unusual for them to restrand themselves," she said, adding DOC staff, volunteers and locals would try to refloat the survivors on Friday afternoon at high tide.

After refloating the first survivors on Thursday afternoon's tide, rescuers saw some of the group head back towards the shore, having swum round the bay towards the open Pacific to the east. By about 2.00 a.m. (1300 GMT) on Friday, the majority of the group had stranded themselves again, and dozens of people returned to the beach to help out. Each of the whales was being attended by two or three people, with damp towels, hoses and containers of water to keep the mammals cool and make sure their blowholes were clear.

The dead whales will be buried in trenches on the beach, Vivequin said, adding that scientists from the Massey University Cetacean Investigation Centre had been offered samples. Local Maori have been offered the bones of the whales for carvings, an important part of their cultural links with nature.

New Zealand's largest ever recorded whale stranding occurred in 1975 on Great Barrier Island when 410 pilot whales went aground.

[http://www.infobeat.com/stories/cgi/]

From the NZ Pilot Whales ...

We are the Pilot Whales. We beached ourselves in New Zealand to call attention to our disastrous plight in the oceans. Swimming in the oceans has become quite deadly, as you know, due to the pollution, fishing nets, whaling ships, and underwater "scientific" experiments.

The experiments done under the water are bringing us to the brink of disaster. The sound waves being generated from experimental equipment is disrupting our equilibrium. We are being bombarded by ELF sound waves that are driving us to beach ourselves as we cannot tolerate the friction/vibration to our bodies. These ELF sound waves interfere with our sonar and communication with each other, and drive us literally "insane". This is a deliberate attempt on the part of your nefarious governments to annihilate us. They know we are here to awaken humanity and bring Light to the Earth, and your governments don't want us interfering with their plans to take over the Earth.

We have no choice but to beach ourselves when we can no longer tolerate the torment created by low frequency sound waves from devices that are used to experiment on us. You, too, above the water are being experimented on, only to a much lesser degree that is so subtle you hardly perceive it.

We are running out of patience and forbearance, and would rather beach ourselves than be slowly tormented to death.

We thank you for publishing our message in your book, as this will bring worldwide attention to us as the "People of the Sea."

Why we beach ourselves / Whaling ships

We wade in the water when it's low tide, and when being followed by trawlers and whaling ships we sometimes beach ourselves. We await anxiously for help to arrive, and when it doesn't, we fight to stay alive. This may sound strange to you, but we prefer to beach ourselves rather than let ourselves be captured by whaling ships. You see, the ships can't go into shallow water, so it's a way that we can usually escape from them instead of outrunning them which often leaves us too tired to continue at such a fast pace. We don't like this solution, but it's one of the few left to us, and one that usually stops the whalers from coming toward us. They know we can outwait them, so they usually leave when we employ this tactic.

You may be wondering about the stories you read about whales beaching themselves for no likely reason. Well, the reason was they were being pursued by whaling ships, and chose to beach themselves rather than be captured alive out in the ocean's depths. It was also a way to bring attention to our plight, if only humans could have understood our reasons for these beachings. There is much you will begin to understand, as you come closer to us in consciousness.

We are your brothers and sisters, and we love you.

Orca. Source: the Internet

We work alongside the Confederation
in many areas and in many ways.
Although our bodies remain in the seas
our consciousness goes
wherever we direct our thoughts.

We are multidimensional,
and while we swim in Earth's oceans,
we are on assignment
in other dimensions simultaneously.
We can be in two places at once,
and often are.

"Attraction," Carlos D. Aleman, Dolphin Art Gallery

Mines on the ocean floor / Multi-dimensional assignment

Today we will talk about the mines that are found on the bottom of the oceans. Know that there are obsolete munition dumps on the floors of our oceans. Although these weapons are now defunct, still they exude dangerous chemicals and litter our oceans.

The ocean is our habitat, and we don't like our "floors" to be cluttered with dangerous weapons, chemicals, and metals. Just as you on land like the freedom to walk in your homes, so do we like to be able to sink to the ocean's bottom and rest there if we wish.

So when the Confederation lands as Earth enters the Photon Belt, one of the first tasks will be a cleanup of our ocean floors. We will oversee this operation, guiding the work crews to specific locations for dematerialization of this clutter.

We work alongside the Confederation in many areas and in many ways. Although our bodies remain in the seas, our consciousness goes wherever we direct our thoughts. For we are multidimensional, and while we swim in Earth's oceans, we are on assignment in other dimensions simultaneously. We can be in two places at once, and often are.

So swim with us during your dream time, and we will show you the wonders of the sea.

**It is hard for us
always dodging the nets
and destructive techniques used to capture us.**

**We Love the Earth
and are sacrificing our lives for it.**

Orca. Source: the Internet

Dodging nets and destructive techniques

We are the Cetaceans living in your oceans. We have found our life to be hard, and yet most rewarding. *For we are the caretakers.* We keep your earth alive with our energies and thoughts, until you can someday be the stewards you were meant to be.

It is hard for us, always dodging the nets and destructive techniques used to capture us. But we have grown adept at avoiding the snares in the oceans and have proliferated in spite of this.

We love the Earth, and are sacrificing our lives for it. We wait for the day when you, too, will recognize the beauty and grandeur of Earth and come to appreciate its uniqueness. Then you will resume your stewardship to her and we will be able to freely communicate with you on land, and our species will join you as *one consciousness* caring for the Earth.

We await this time when *all species will live together as one.*

Killing and mutilating other species has no place in God's plan

We greet you in the Light of the *one Creator of All That Is*. We, too, are creatures of God and we, too, have our hopes and dreams for life that we would like to see fulfilled. All creatures on Earth, whether large or small, have their life plans set out before them when they come to Earth, and all strive to accomplish their plans before leaving this planet. For this is a planet of Free Will, and we can choose which aspects of ourselves we wish to enrich, without the encroachment of others getting in our way. This is how it was meant to be. However, mankind has interfered in our lives to such an extent that we can no longer hope to achieve the results of our vision.

Killing and mutilating other species has no place in God's plan, and is a transgression of Universal Law. The sooner humanity sees this, the sooner all life on Earth will be safe to evolve naturally and in harmony with the rest of the universe.

So we ask you, Earth humans, to look at your ways of living, and honor only that which prolongs and protects all life species on Earth. For without all species in harmony, life as you know it will be doomed to a pathetic existence of survival, until its Light goes out completely, leaving a barren and hostile landscape behind.

"Peaceful Moments," courtesy of Janet Biondi

The Whale's Perspective on Whaling

This is the Whale's response to an article entitled "Save the Whalers" printed in the Wall Street Journal (9/9/97) by Mr. Aron, a former U.S. Commissioner of the International Whaling Commission. It is about pressuring the U.S. delegation to the International Whaling Commission to lift the moratorium on whaling, and defend Japan and Norway's slaughter of whales. Mr. Aron states that "whales and whalers can co-exist."

The Whales Respond ...

We are the Whales, who speak to you now, your Earth-bound brothers and sisters residing in Earth's oceans. Do you know of "interspecies communication," or do you only know how to kill other species instead of communicating with them? We challenge you to listen to us and to intermingle with us, but not to murder us.

It is true that "whales and whalers can co-exist," but not if we're all dead, murdered by whaling fleets and aboriginal tribes in search of a livelihood. Surely there are more ways to earn a livelihood than by hunting down another intelligent species on your planet!

We whales exist to balance the Earth's ecosystem, and live family lives and spiritual lives to progress and evolve as a race of sentient Beings just as you humans do. We don't harvest humans, even though you have proliferated to such an extent that you have almost taken over the globe. Should we come onto your land masses and harpoon you to death so that we in the seas can live? We certainly wouldn't process your bodies and eat you as food, as you have no nutritional value, and spiritually you act as if you're dead. Look how you've decimated the land and exterminated countless other species. And now you're still intent on forever exterminating us. Don't you ever learn from your wanton destruction of God's creatures that all are sacred and all are needed in order for life on land to exist?

The article refers to us as a resource to be harvested. How can you put us in the same category as a field of grain or a kelp bed? What authority does humanity have to declare us as "harvest"? We do not "belong" to you as do cattle and pets. We are a free species, trying to live and do our work, just like you. So please leave us alone!

All life has consciousness, and we whales are highly evolved conscious Beings, here on Earth to help you humans evolve into a higher state of consciousness, in which you won't kill other species for a "livelihood," but will know how to co-exist in perfect harmony with all species. This way, you could earn a livelihood from caring for the Earth instead of destroying it.

The article also states that, "whales are unquestionably extraordinary animals. While they possess a certain intelligence, however, there are no data to support the belief that they are at or even near the top of the animal intelligence scale." How are you measuring intelligence? Just because we cannot pass your exams doesn't make us any less than you. Our intelligence is so far removed from yours that you cannot begin to appreciate it.

You depend on us in ways you couldn't even imagine, and we are all bound in a living network of consciousness. If you persist in killing us, you will doom yourselves into rapid extinction, as all species are needed to exist in order for Earth to continue as a home for ALL life, including humans.

We are aware of something that you have forgotten: that the primary identity of human and cetacean is that of *soul*. These bodies we both occupy for a few short years are simply costumes. Behind the disguises, *we are the same*. So let our only contact be recreational, and let us co-exist peacefully on this beautiful planet.

Slaughtered 40 foot Sperm Whale. Photo by Jeff J, Mitchell, courtesy of Whales On The Net. [www.whales.magna.com.au] email: gclarke@magna.com.au

Pilot Whale Slaughter

Every year in the Faroe Islands about 1500 Pilot Whales are brutally slaughtered. While still alive, these pilot whales are brutally impaled with steel hooks and dragged ashore. The Faroese also kill other small whales, dolphins and porpoises. These gentle creatures are herded into bays and killed by islanders armed with 6-inch whaling knives and heavy steel hooks, known as gaffs.

First the gaff is hammered into the whales flesh, often causing deep wounds and sometimes even puncturing the blowhole. Then as the animal thrashes in pain, it is dragged by a rope attached to the gaff towards the shore. The islanders slice down just behind the blowhole with a knife, trying to sever the main arteries and spinal cord. This may take a couple of minutes. But if the whale struggles too much, or the killer is inexperienced (or even drunk), death may take much longer. No whales are spared from this agony. Whole families, including pregnant mothers, are slaughtered in just a few hours.

Text and photo courtesy of Earthisland
(http://www.earthisland.org/immp)

Norway's Rampant Whale Slaughter

1995 Report from the International Marine Mammal Project (IMMP)

Renegade Norwegian whalers may soon be joined by Icelandic and Japanese whalers in defying the International Whaling Commission (IWC) whaling ban. In addition to minke whales that Norway continues to kill in defiance of the IWC's moratorium, reports now indicate that endangered fin and humpbacks are also being slaughtered illegally. But the Clinton Administration refuses to impose economic sanctions on Norway.

Our project continues to press for an end to commercial whaling. By educating the public about the threats still facing whales, urging the Clinton Administration to ban all fish imports from Norway, as well as conducting undercover investigations of illegally smuggled whale meat, we aim for a final end to the unnecessary whale slaughter.

In 1993, Norway openly defied an IWC global ban on commercial whaling and killed 226 minke whales in the Northeastern Atlantic Ocean. In response, the US Commerce Department certified that Norway was "diminishing the effectiveness" of the IWC and thereby authorized President Clinton to ban the import of certain Norwegian products. However, last October, Clinton notified Congress that he would not impose sanctions until "good faith" negotiations to persuade Norway to abide by the IWC commercial whaling ban had concluded. But, no good faith negotiations ever took place between the US and Norway. Instead, Vice President Gore held secret meetings with Norwegian Prime Minister Gro Brundtland and helped Norway obtain a whale kill quota from the IWC at its annual meeting in May. Once news of these meetings was leaked, Earth Island and a host of other environmental organizations waged a campaign to alert the public to the administration's sell out. Full-page newspaper advertisements criticized the US' actions, and citizens sent thousands of letters and telegrams to the White House to pressure the administration to protect the whales.

At the IWC meeting, Norway failed in its attempt to overturn the global whaling ban. No country was granted a quota for commercial kills. In addition, a resolution was passed establishing a whale sanctuary in the Southern Ocean of Antarctica. Commercial whaling is prohibited in the sanctuary which covers one-third of the world's oceans and contains close to 90 percent of the remaining whales. EII Executive Director David Phillips called the sanctuary, "... a big step forward for international whale conservation."

Even with the establishment of the sanctuary and the IWC's refusal to grant any commercial quotas, Norway announced that it would again defy the moratorium. Since May 1994, Norway has killed 301 minke whales, a 20 percent increase over its kill last year. In addition, Japan responded by filing a formal objection to the Antarctic Sanctuary, saying that the minke whale population is large enough in that region to sustain commercial whaling. In 1994, Japan proceeded to kill 330 minke whales, including 21 in the North Pacific Ocean under the guise of "scientific research." Iceland has also stated its intention to commence killing whales again by the spring of 1995.

Up to this point, Norway has defied the IWC commercial whaling ban with impunity. Its recent vote not to join the European Community (EC) makes the EC's ban against commercial whaling inconsequential to Norway. The EC is now unable to pressure Norway to stop whaling; it could embargo Norwegian products. It is illegal for whale meat to be bought or sold in any EC nation.

While Norway generates up to $7 million annually by selling whale meat, a US ban on its fish imports would cost Norway more than $100 million in lost revenue. The Clinton Administration could take the lead and impose sanctions against Norway, sending a strong signal that Norway's blatant disregard for international bans cannot be tolerated. Norway's whale slaughter affects not only the minke whale. It opens markets to the killing of all types of whales, including those populations on the brink of extinction. In an attempt to assess the illegal trade in whale meat, IMMP is supporting a team of undercover investigators to take DNA samples of whale meat sold around the world. "It is quite clear that meat from highly endangered fin and humpback whales is on market shelves right now," said Phillips.

The IWC meets again June 1995 in Ireland. We will keep up the pressure to ensure that the ban stays in place and is properly enforced so that the world's whales will no longer be killed illegally.

IMMP responds to Norwegian whaling claims

Because of the international pressures being put on Norway to end its yearly minke hunts, the Norwegian Ministry of Foreign Affairs has produced a pamphlet (Some Questions About Norwegian Minke Whaling) in which they make a number of claims to try and support the hunts:

1. Norway claims that the Scientific Committee of the International Whaling Commission (IWC) has unanimously concluded that this stock (the minke whale) almost certainly numbers between 61,000 and 117,000 animals with the best available estimate being 87,000 animals. IMMP notes that, not only have environmental groups questioned these figures, but members of the IWC have also questioned the figures. The environmental community, and some IWC members, question the Norwegian whale population estimates because all research and calculations were done by Norwegian officials without any review from outside scientists.

2. Until recently, Norway has consistently used the 87,000 number to justify its hunts. They claimed that because of the figures it was clear that the stock was large enough to support a limited harvest. IMMP reports, however, that startling new evidence has surfaced that Norway's whale population estimates have, in fact, been grossly exaggerated and that its formulas for calculating populations contain numerous errors. A special meeting of the IWC's Scientific Committee recently examined the data, and even Norway's own scientists have been forced to admit the errors. In April, the Norwegian government reduced its estimate of the minke whale population by more than 20 percent to 69,600. Further review may show that even this estimate is far too high. Also, because of this revision, the Fisheries Ministry has reduced this year's whale kill quota from 301 to 232 whales. In fact, only 30,000 whales may exist which would grant a zero quota.

3. Norway claims that its whaling is carried out in accordance with the international agreements that are relevant to whaling. These are the Convention on the Law of the Sea and the International Convention for the Regulation of Whaling, which sets

the framework for the activities of the IWC. They go on to say that Norway plays a leading role in binding international cooperation on environmental protection and resource management. IMMP points out, however, that they are in fact directly contradicting this claim as there has been, and still exists, a global moratorium on whaling. Since the Norwegian government has allowed whale hunts for the past two years, they have in fact been violating an international environmental agreement. Also, now that Norway has defied the international community on this issue, they have let the proverbial genie out of the bottle. Norwegian whalers now say that unless they receive full financial compensation of $8,500 (1,565 kroner) per whale for the reduction in the planned quota, they will kill all 301 whales, in defiance of both the IWC ban and the Norwegian government limits.

4. Finally, Norway claims that the most important argument in favor of whaling is the principle that states have a right to utilize their renewable natural resources on a scientific basis. Nations have a right to determine their own affairs, but not in violation of international treaties that they have signed on to. If Norway is permitted to violate the commercial whaling ban and base their catch on inaccurate whale population data, there will be little to prevent Japan, Iceland, and a host of other countries from doing the same. This precedent could cause the entire IWC moratorium to crumble. It would also set a dangerous precedent globally if countries are allowed to back out of environmental treaties and agreements if they just don't feel like honoring them anymore.

Voice Your Opinion

- Encourage local newspapers, radio, and TV stations to cover the tuna/dolphin and whaling issues.
- Write articles for your school newspapers or business publications to educate others about the tuna/dolphin and whaling issues.
- Send letters to your senators and representatives encouraging them to take a strong position against commercial whaling.
- Write letters to the editor of your local newspapers and express your concern for marine mammals.

Write Letters

Write to Pres. Clinton and VP Gore. Demand that Clinton place sanctions on Norway until it stops commercial whaling.
>The White House, 1600 Pennsylvania Ave., NW
>Washington, DC 20500

Write to Gro Harlem Brundtland, Prime Minister of Norway, and demand that Norway put an end to the senseless slaughter of whales.
>Gro Harlem Brundtland
>Prime Minister of Norway
>Norwegian Embassy
>2720 34th St., NW
>Washington, DC 20008

Contact Whales Alive to learn more about their efforts to protect whales and how you can get involved in their work.
>Whales Alive
>PO Box 2058
>Kihei, HI 96753

Urge the Marriott Corporation to adopt an EII approved, dolphin-safe corporate policy for its many hotels and restaurants.
>Mr. Richard Marriott, Vice Chair
>Marriott Corporation
>Marriott Drive
>Washington, DC 20058

Contact Brenda Killian for information about the International Monitoring Program.
>International Monitoring Program
>Earth Island Institute
>2439 South Kihei Road, Suite 202B
>Kihei, HI 96753

Write to the Free Willy/Keiko office for updates about Keiko and his rehabilitation.
>Free Willy/Keiko Foundation
>Earth Island Institute
>300 Broadway, Suite 28
>San Francisco, CA 94133
>Email: marinemammal@igc.apc.org

Ocean Alert is an Earth Island Institute publication. Visit their website at: http://www.earthisland.org/immp

Urge the InterAmerican Tropical Tuna Commission (IATTC) to set the yearly dolphin kill rate for the international fleet to zero.

> IATTC
> Scripps Institute of Oceanography
> 8604 La Jolla Shores Dr.
> La Jolla, CA 92037

Write to the National Marine Fisheries Service (NMFS) and demand that it enforces the MMPA to the fullest extent of the law.

> Dr. William Fox
> NMFS
> 1335 E. West Hwy.
> Silver Springs, MD 20910

Part Six

Messages to Green Peace

Our Dear Brothers and Sisters in Green Peace:

We are the Cetaceans, channeling messages through Dianne Robbins, our channel on land.

We wait for the time when we can be in direct telepathic contact with you. Although some of your crew members out at sea are telepathic, we wish to send information directly to you through our channel, in hopes of forming a tighter alliance with those countries who are still hunting our species. Our species can no longer tolerate these hunts on our population, as all of us are required to be alive to continue our caretakership of Earth.

The massive killings have decimated our stock, and many of us have left the planet on our own. We feel that we have every right to live in peace in the oceans, just as you do on land.

This letter is to bring to your attention the information that we have telepathically sent to Dianne in the form of messages that we hope you will include in your newsletters to Green Peace members and to countries that still disregard the treaties that have been established to protect our species.

With all our love to you, for all you do for us.

The Cetaceans, July 5, 1995

Green Peace / Establish an alliance with Earth humans

We are the Cetaceans greeting you on this summer day when the oceans swell with water and the current is swift. Know that all life in the sea knows about the upcoming Photon Belt, and we are all tuned into Divine Consciousness. We are all aware of Earth's events, *always*, since we, too, are a part of the Divine Plan in which all species are involved, no matter where they are. So we, too, await the Photon Belt, just as you do.

We gather our pods together for a great assembly at this time of year, just as you gather together in large groups on land. We take a counting, and plan for events that we see occurring in the immediate future. We are all connected to one another through our hearts, and we know immediately when one of us is in trouble or has some difficulty. This way, others of us can come to the rescue in a matter of minutes. We are wary of all predators whose aim is to shorten our life span, and use our carcasses for their products. We try to avoid the whaling ships at all cost, but sometimes we don't see them as we gaily glide along through the ocean, enjoying the sweet air and communicating with all around us. *All products for consumer use can be found in the Earth. Just look to nature and you see that all you need is already provided for you in abundance.*

It isn't necessary to take anyone's life to sustain your consumption or to meet your needs. *The Earth is a great Mother, always providing everything in a clean, pure form.* It is only humans who adulterate nature's bounty causing sickness, disease, and death.

Hopefully, we will be able to establish an alliance with Earth humans in the near future, so that our safety is no longer threatened. This has been a goal for us for eons, and we are now close to implementing it with the help of Green Peace. This humanitarian organization has done much to assuage our hunters, and to them we are most grateful. Many of these volunteer crew members are themselves telepathic and can communicate with us. It is to these ones we give information as to our whereabouts and guide their efforts to eliminate the whaling "business."

Someday you will all realize how easy life can be by following nature's laws and nature's way of life. Then you will know bliss, for all will flow to you naturally and easily, provided by the Great Force of Life itself. We wait patiently for the great ascension waves to carry us all back home again to the safety of the Higher Dimensions where all our family await us.

Bon Voyage, our mates on land.

"Attraction" by Carlos D. Aleman, Dolphin Art Gallery

Green Peace instrumental in our survival /
Our scouts are on lookout for whaling ships

We are the Cetaceans grouped together in the middle of the sea and sending our thoughts out to you sitting at your desk. We, too, once sat down, when we were land bound mammals. Now we just float instead and relax as the currents move us to and fro.

We are a simple species, yet very complex in our mechanism for survival. Our simple life is much sought after by many on land, and yet we are complex by nature. Because we have our full consciousness, we can operate on many different levels at once, knowing and participating in all these different functions at the same time. This is indeed a complex operation on our part, and yet it is so simple to achieve.

Our lives are fraught with danger, and yet we are able to sing, dance, play, and enjoy our "life style" even amidst the dangers. Our scouts are constantly on the lookout for humans in whaling and fishing ships, and when sighted, the call goes out by sonar to all in the ocean. Should any one of us incur danger, our brothers and sisters are aware of it and close in immediately to help us regain our position.

Know that Green Peace has been instrumental in our survival, saving hundreds of thousands of us out at sea. We wish to work more closely with Green Peace, and relay information to them through you. We hope they will agree to receiving our messages, as we wish to have more direct contact with their operation out at sea.

We are grateful to them for their commitment to our species, and for their work in bringing peace between us. All we wish for is a peaceful alliance with humans and the God-earned right to safely navigate through our oceans without harm befalling any of us. So we call on you, our brothers and sisters in Green Peace, and urge you to continue your peaceful ways to bring peace to Earth's oceans.

On behalf of all of us out at sea, we wish you God's grace and God's love for all you do for Earth's mighty population of cetaceans. We are ONE in Love.

Part Seven

Blue Whale Gathering,

August 1996

Blue Whale

Blue Whale Gathering, August 1996

The following quote was taken from CNN news off the Internet:

"Blue Whales are rare...but as many as a hundred of them are congregating off the coast of Santa Barbara at the Channel Islands Marine Sanctuary."

I then asked the one group mind of Earth's Cetaceans to tell me what this gathering was about. Here is their reply:

We hear your thoughts about our Great Gathering off the California coast a few weeks ago. Yes, it was a planned gathering and all the Blue Whales from all the oceans were invited to participate in this great conclave where we discussed the plans for our eventual movement into the higher frequencies of Light.

It was indeed a very special occasion, as we haven't had a gathering like this in millennia, where we were in eyesight of humans on shore. The energies are quickening now, and our group helped to ground the higher frequencies into the ocean, where they will then radiate out to the land.

We had many things to talk about, and it was an enriching and grand meeting. We talked about our survival out at sea, and your (humans) survival on land and how fragile the line between the two are. The strength of one depends on the strength of the other.

We whales are not safe in the seas, as the pollution and deadly devices are attacking us as we swim through our own homeland, without protection. We talked about the risks we are all taking by being here in Earth's oceans, and how we all yearn for our home in the stars, where it is safe and clean, and where we are all respected and admired for our evolutionary growth.

We have had many meetings and gatherings in the oceans

Greetings from Earth's oceans. Today it is sunny. We float on the currents as the waves toss us to and fro. We love this time of year (August), as everything is in bloom and all of life is bursting with hope. We love our homes in the sea. They are complete and protective from earth humans, except of course for the whaling boats and nets that infringe upon and intrude into our home environment.

We are a joyous folk, very much like your brothers and sisters in the subterranean cities. We, too, have had the luxury to evolve in consciousness as our oceans have given us the privacy and protection necessary for our evolution. We have been fortunate in this respect. Since humans "discovered" us as a product for consumption, we have lost the privacy and protection offered by the Earth's vast oceans.

We have had many meetings and gatherings in the oceans, to discuss our plight with whaling vessels and fishing nets, and to discuss our Earth's voyage through the Photon Belt. These meetings have been privately attended, away from human eyes, lest we be attacked while all assembled together. It was safer for us when we gathered near Santa Barbara in the Pacific (August 1996), since in that portion of the ocean we are fully protected from trawlers and fishing nets. We are very wise, having accumulated ages of wisdom, and we use our wisdom, collectively, to help Earth spin into the Golden Light and shed her cloak of density.

Part Eight

The Great Awakening

The Great Awakening
is now taking place
on planet Earth.

"*Memories,*" by Carlos D. Aleman, Dolphin Art Gallery

In Tribute To My Daughter

To Helen, our Dear Sister of Light: We greet you at the dawn of a New Age on Earth. This is the Age of Aquarius, a time when many prophesies are coming to an end. It is a glorious time to be on Earth, for we are seeing the awakening of millions of people to their identity. People are awakening to who they are and why they're here on Earth.

You, too, will be awakening to the wonders of the Universe, and you, too, will begin questioning your identity and purpose for being here on Earth. This is the "Great Awakening" and it is taking place *now*—right here on planet Earth. All life is awakening to the magic and wonderment of life, and all life is being guided on its path toward the ever increasing amounts of Light that are flowing to Earth in a spiral of dazzling bursts of stars exploding all around you.

So, come swim with us at night, and we will guide you on your path to the stars. We are the Cetaceans and we love you.

In Tribute to My Son

To Jason, our Dear Brother on land: We are the Cetaceans, sending our love to you from the depths of the oceans. We swim with you at night in your dream state, when your soul frees itself from your body to traverse the Universe.

You are a great Being of Light, who has come to Earth to help lift Earth and humanity into higher states of consciousness. You are a very evolved soul, whose specialty is broadcasting love.

Now is the time to open to the wonders of the Universe around you. For the Universe is not all you perceive it to be from the third-dimensional plane of existence. Instead, contrary to your five senses, the universe is vast and exists on many levels of understanding and perception. As you open to these multiple levels of perception, you will know and understand more about who you are and why you are here on Earth. For we have all come here with a purpose, and now is the time to get in touch with that purpose and to explore all life through our multi-dimensional senses.

We are here to guide you and to assist you on your journey through life. We are your brothers and sisters from the depths of the oceans, waiting for you to join us in consciousness, even though we are at sea and you are living on land. Neither the land nor the sea is a barrier, for our thoughts transverse all matter, and connect us in consciousness as ONE, no matter where we are or what we are doing. So contact us in your thoughts, as we are always beaming our love out to you on land.

We await your contact. We love you.

— *The Cetaceans*

Part Nine

We Salute You In The Light

Orca. Source: the Internet

Freedom is in oneness

We are aligned with you on the currents of Light winding around our planet. The currents ebb and flow like the water, and carry our signals to shore where humanity can connect with them.

So, as more and more of you on land connect to us at sea, the day will dawn when we'll all be free. For freedom is in *oneness*. *Freedom is knowing that we are all one, no matter what our life form is.* For these life forms that we inhabit are merely shields to protect and to keep us grounded in third-dimensional life. Our bodies offer us the opportunities to evolve ourselves, through pretending we are separate. This allows our consciousness to concentrate on the reality we're presently in, without becoming confused by the other dimensions surrounding us.

So hold the Light and hold us in your thoughts, for it is your thoughts that bring us together as one—one Light shining from off shore and connecting to ALL.

We long for your company

Today it is windy and the sea is choppy. We are gathered around you in our etheric bodies as we speak with one voice from Earth's seas. We spread our consciousness around you, and envelop you in our souls. For we are one soul although our forms are different. This difference appears on all planets in all solar systems since the creative force is ever experimenting and ever expanding in consciousness and in form. For life is *all one*, no matter how we are or where we are. That's the Law of Creation, and it's the binding force of the Universe. It's the glue that holds all of us together, for we're all linked in *one* consciousness to *one* creation, no matter who we are, where we are, or what we look like. It's that indissoluble bond that's everywhere and in everything. Some call it the Life Force that's in all of us—the same life force all over the world, regardless of our differences.

So today the waves are churning and the water is choppy and the wind is blowing. We are diligently moving along engrossed in our day's tasks for the Confederation. There isn't much to report because, on days like this, there's only us out in the sea, since the whaling boats don't venture out this far. So although the water is somewhat difficult to deal with, it at least affords us the temporary safety we permanently long for.

So come with us as we travel through these crescent waves making our journey through Earth's oceans. We are the Cetaceans and we long for your company.

We swim with humans who are lost at sea

We are the Cetaceans. We are grouped together as we broadcast this message to you. We swim in the seas and bask in the sun. We are friendly and jovial, and welcome all humans who enter our waters. *For the oceans are for all, just as your land is for all.* No one is barred from entering our oceans. We often swim with the humans who are floundering out at sea, until help arrives or until we can guide them to shore.

We take great pride in our prowess and strength, and great responsibility when someone is lost out at sea. Many times we are able to telepathically contact humans who are adrift and give them information that will guide them back to land.

There are so many ways that we could work with you, if you would but recognize us as equal to you. That is the way it was supposed to be originally, when the Earth was populated with so many different species. We all had our own roles to play, and we all had our roles to play with each other.

This was the perfect plan, had humans not deviated from it. It was the Divine Plan that would have led us all in a much different direction than we now find ourselves. We could have reached the summit of glory long ago, instead of reaching the bottom of degradation in which we now find our Earth.

So let's begin now and enter into a partnership that will bring us and all Earth to greater heights of understanding and greater heights of glory than we now find ourselves in. *It is not too late.* We await your call, and will work with you at night in your dream state.

Farewell, our brothers and sisters on land.

We are grateful to all on Earth
who have been vociferous
in protecting our species.

For it is through the light of your voices
that we are still here
and still able to carry out our tasks
as caretakers of this planet.

On behalf of all of us out at sea,
we wish you God's grace
and God's love for all you do
for Earth's mighty population of Cetaceans.

We Salute you in the Light.

Orca. Source: the Internet

Feel your heart as it beats.
Listen to your pulse as it throbs to the rhythm of life.
Feel your blood as it courses through your body,
carrying oxygen and nutrients to every cell.
You are a self-contained storehouse of life.
You are a microcosm of the Universe.

You are a replica of God.
All of life pulses through you.
All knowledge is contained within you.
All can be accessed by you.
Just go within and ask, listen, and feel,
and then you will know.
All answers are within you,
for you are within ALL,
and ALL is ONE.

— *Dianne Robbins*

Humpback. Source unknown

About the Author

Dianne Robbins is a telepath and empath, so not only does she receive information but also feels what Corky, Keiko, and the other Cetaceans feel. She is an Interspecies Communicator, specializing in raising people's awareness of the continuing inhumane killing of Cetaceans, and the imprisonment of their species. She is of Cetacean lineage and consciousness, and has been a telepathic channel for the Cetaceans in other lifetimes. Channeling under the name LAILEL, today, she is one of the Cetaceans' voices to the world.

Dianne has always been connected to the Cetaceans, and was an active member of Green Peace in the 1970s, only at that time, she didn't realize that she could hear them. Being telepathically linked to the Cetaceans in previous lifetimes, it was easy for her to connect with them now, once she realized that she could do this. They had been waiting for some time for her to realize this, and planted many signs along her path to call her attention to them.

Dianne started her search in the late 1980s, with the questions: "who am I, why am I here, and what am I supposed to be doing?" She spent hours each day for many years in meditation and silence, out in nature, walking, and just sitting quietly to find the answers to her questions, and started receiving telepathic messages from the Cetaceans in the early 1990s. Many of the messages in this book were channeled while Dianne was at Deerfield Beach, FL. She would sit alone by the ocean with pen and notebook, and telepathically call to the Cetaceans, "I am ready." They would then dictate their messages to her.

Dianne is irresistibly drawn to the ocean, visiting the beach every day. She yearns to live near the ocean, and dreams of moving to Florida where her family and many Cetaceans are. Family life is very important to her, and she misses her family in Florida from whom she is separated, just as Corky and Keiko are separated from their families—only she has the freedom to visit her family as she wishes. Dianne fervently hopes that Corky and Keiko, and all Cetaceans in captivity will be freed, so that they, too, can be with their families again; and that someday she, too, can be with hers.

Dianne can be reached at PO Box 10945, Rochester, NY 14610, and telos@rochester.rr.com - Phone: 585-442-4437

TGS Publishers

22241 Pinedale Lane
Frankston, Texas 75763

HiddenMysteries.com
903-876-3256